T0139827

Advanced Multicore Systems-On-Chip

Abderazek Ben Abdallah

Advanced Multicore Systems-On-Chip

Architecture, On-Chip Network, Design

 Springer

Abderazek Ben Abdallah
School of Computer Science
  and Engineering
The University of Aizu
Aizu-Wakamatsu, Fukushima
Japan

ISBN 978-981-13-5565-3   ISBN 978-981-10-6092-2   (eBook)
DOI 10.1007/978-981-10-6092-2

Printed on acid-free paper

This Springer imprint is published by Springer Nature
The registered company is Springer Nature Singapore Pte Ltd.
The registered company address is: 152 Beach Road, #21-01/04 Gateway East, Singapore 189721, Singapore

To Sonia, Tesnim and Beyram.

Preface

Nowadays, the technology has become an essential pawn in our life that is not restricted anymore to academic research or critical missions; but it is moving away to provide the simplest and easiest services that we need or desire for our daily life. With the expanse of technology and the rising of new trends every day, the necessity to process information anywhere and anytime is becoming the main goal of developers and manufacturers.

Systems on chip (SoCs) are embedded systems composed of several modules (processors, memories, input/output peripherals, etc.) on a single chip. With SoCs, it is now possible to process information and execute critical tasks at higher speed and lower power on a tiny chip. This is due to the increasing number of transistors that can be embedded on a single chip, which keeps doubling approximately every 2 years as Intel co-founder Gordon Moore predicted in 1965. This made shrinking the chip size while maintaining high performance possible. This technology scaling has allowed SoCs to grow continuously in component count and complexity and evolve to systems with many processors embedded on a single SoC. With such high integration level available, the development of multi and many cores on a single die has become possible.

Historically, the SoCs paradigm has evolved from fairly simple unicore single memory designs to complex homogeneous/heterogeneous multicore SoC (MCSoC) systems consisting of a large number of intellectual property (IP) cores on the same silicon. To meet the challenges arising from high computational demands posed by latest state-of-the-art embedded and consumer electronic devices, most current systems are based on such paradigm, which represents a real revolution in many aspects of computing.

The attraction of multicore processing for power reduction is compelling in embedded and in general purpose computing. By splitting a set of tasks among multiple cores, the operating frequency necessary for each core can be reduced, thereby facilitating a reduction in the voltage on each core. As dynamic power is proportional to the frequency and to the square of the voltage, we are able to obtain a sizable gain, even though we may have more cores running.

As more and more cores are integrated into these designs to share the ever increasing processing load, the primary challenges are geared toward efficient memory hierarchy, scalable system interconnect, new programming models, and efficient integration methodology for connecting such heterogeneous cores into a single system capable of leveraging their individual flexibility.

Current design methods are inclined toward mixed hardware/software (HW/SW) co-designs, targeting multicore SoCs for application specific domains. To decide on the lowest cost mix of cores, designers must iteratively map the device's functionality to a particular HW/SW partition and target architectures. In addition, to connect the heterogeneous cores, the architecture requires high performance-based complex communication architectures and efficient communication protocols, such as hierarchical bus, point-to-point connection, or the recent new interconnection paradigm—network on chip.

Software development also becomes far more complex due to the difficulties in breaking a single processing task into multiple parts that could be processed separately and then reassembled later. This reflects the fact that certain processor jobs could not possibly be easily parallelized to run concurrently on multiple processing cores and that load balancing between processing cores—especially heterogeneous cores—is extremely difficult.

This book is organized into nine chapters. The book stands independent and we have made every attempt to make each chapter self-contained as well.

Chapter 1 introduces multicore systems on chip (MCSoCs) architectures and explores SoCs technology and the challenges it presents to organizations and developers building next-generation multicore SoCs-based systems.

Understanding the technological landscape and design methods in some level of details are very important. This is because so many design decisions in multicore architecture today are guided by the impact of the technology. Chapter 2 presents design challenges and conventional design methods of MCSoCs. It also describes a so-called scalable core-based method for systematic design environment of application specific heterogeneous multicore SoC architectures. The architecture design used in conventional methods of multicore SoCs and custom multiprocessor architectures are not flexible enough to meet the requirements of different application domains and not scalable enough to meet different computation needs and different complexities of various applications. Therefore, designers should be aware of existing design methods and also be ready to innovate or adapt appropriate design methods for individual target platform.

Understanding the software and hardware building blocks and the computation power of individual components in these complex MCSoCs is necessary for designing power-, performance-, and cost-efficient systems. Chapter 3 describes in details the architectures and functions of the main building blocks that are used to build such complex multicore SoCs. Readers with a relevant background in multicore SoC building blocks could effectively skip some of the materials mentioned in this chapter. The knowledge of these aspects is not an absolute requirement for understanding the rest of the book, but it does help novice students or beginners to get a glimpse of the big picture of a heterogeneous or homogeneous MCSoC organization.

Whether homogeneous, heterogeneous, or hybrid multicore SoCs, IP cores must be connected in a high-performance, scalable, and flexible manner. The emerging technology that targets such connections is called an on-chip interconnection network, also known as a network on chip (NoC), and the philosophy behind the emergence of such innovation has been summarized by William Dally at Stanford University as *route packets, not wires*.

Chapters 4–6 presents fundamental and advanced on-chip interconnection network technologies for multi- and many-core SoCs. These three chapters are all very important part of the book since they allow the reader to understand what needed microarchitecture for on-chip routers and network interfaces are essential towards meeting latency, area, and power constraints. Reader will also understand practical issues about what system architecture (topology, routing, flow control, NI, and 3D integration) is most suited for these on-chip networks.

With the rise of multicore and many-core systems, concurrency becomes a major issue in the daily life of a programmer. Thus, compiler and software development tools will be critical towards helping programmers create high-performance software. Programmers should make sure that their parallelized program codes would not cause race condition, memory-access deadlocks, or other faults that may crash their entire systems. Chapter 7 describes a novel parallelizing compiler design for high-performance computing.

Power dissipation continues to be a primary design constraint and concern in single and multicore systems. Increasing power consumption not only results in increasing energy costs, but also results in high die temperatures that affect chip reliability, performance, and packaging cost. Chapter 8 provides a detailed investigation of power reduction techniques for multicore SoC at components and network levels. Energy conservation has been largely considered in the hardware design, in general and also in embedded multicore system's components, such as CPUs, disks, displays, memories, and so on. Significant additional power savings could be also achieved by incorporating low power methods into the design of network protocols used for data communication (audio, video, etc.).

Chapter 9 ties together previous chapters and presents a real embedded multicore SoC system design targeted for elderly health monitoring. For this book, we used our experience to illustrate the complete design flow for a multicore SoC running an electrocardiogram (ECG) application in parallel. Thanks to the recent technological advances in wireless networking, embedded microelectronics, and the Internet, computer and biomedical scientists are now capable to fundamentally modernize and change the way health care services are deployed. Discussions on how to design the algorithms, architecture, register-transfer level implementation, and FPGA prototyping and validation for ECG processing are presented in details.

This book took nearly 2 years to complete. It evolved from our first book and is derived from our teaching experiences in embedded system designs and architecture to both undergraduate and graduate students. Multicore systems paradigm created stupendous opportunities to increase overall system performance, but also created many design challenges that designers must now overcome. Thus we must continue innovating new algorithms and techniques to solve these challenges.

The author is thankful to numerous colleagues and graduate students for their lively discussions and their help in preparing the manuscript of this book. Special thanks are due to the publishers in bringing out this book quickly, yet maintaining very high quality.

Aizu-Wakamatsu, Japan   Abderazek Ben Abdallah

Contents

List of Figures

List of Tables

Chapter 1
Introduction to Multicore Systems On-Chip

Abstract Systems On-Chip (SoCs) designs have evolved from fairly simple unicore, single memory designs to complex heterogeneous multicore SoC architectures consisting of large number of IP blocks on the same silicon. To meet high computational demands posed by latest consumer electronic devices, most current systems are based on such paradigm, which represents a real revolution in many aspects in computing. This chapter presents a general introduction to the multicore System-On-Chip (MCSoCs). We start this chapter by describing the needs for multicore systems by today's general and embedded application domains. Design challenges and basics multicore SoCs hardware and software design are also described.

1.1 The Multicore Revolution

The major chip manufacturers and processor architects have historically invested time and money in micro-architectural and performance enhancements. Many of these efforts such as deep pipelining, increased large cache size, and sophisticated dynamic ILP (Instruction Level Parallelism) extraction exhibit diminishing returns due to increased area and power consumption. When considering the limitations associated with voltage supply scaling, threshold scaling, and clock frequency scaling, along with the above design complexity, architects were already looking for an alternative to the single-core approach. Multicore was therefore the natural next revolution in staying on the ever increasing performance driven curve. But, was it really the good timing to switch from uni-processor approach to the more complex parallel structure of multiprocessor/multicore platforms? The direct answer from major hardware companies was very clear: yes; it is time for revolution and not for evolution! This important decision was fueled by the shift that started from around 2004 when market leaders in the production of general purpose computer systems and embedded devices started offering an increasing number of cores (processors), in which multiple cores communicate directly through shared hardware caches, providing high concurrency instead of high clock speed. This shift contributed to an unprecedented paradigm that has led to what is know today as multicore revolution.

© Springer Nature Singapore Pte Ltd. 2017
A. Ben Abdallah, *Advanced Multicore Systems-On-Chip*,
DOI 10.1007/978-981-10-6092-2_1

The main reason behind this shift can be simply explained by the limits of process technologies. As the computing needs of each processor type grew year by year, the traditional response by the semiconductor industries was to increase the clock frequency of the processor core. However, as processor frequencies increase, other issues such as power consumption, thermal power, the inability to find sufficient parallelism in the program and lagging memory bandwidth become real obstacles to further advancements.

In a typical multicore SoC system, a single physical chip integrates various components together. The single chip may contains digital, analog, mixed-signal, and often radio-frequency functions. Further, each individual core can run at a lower speed, which reduces overall power consumption as well as heat generation. For example, Intel Polaris multicore chip contains 80 cores, each containing two programmable floating point engines and one five-port messaging passing router [1]. This integration approach offers significant price, performance, and flexibility over higher speed single-core processor design.

1.1.1   The Impact of Moore's Law

One of the guiding principles of computer architecture is known as Moore's Law. In April 1965, Gordon Moore wrote an article for Electronics magazine titled [Cramming more components onto integrated circuits] [2]. He predicted that the number of transistors on a chip would double every 12 months into the near future. Although this exponential trend has gradually lessen to doubling transistors every 18 months, it remains the driving force behind the integrated circuits industry. This law over the years has provided a road-map for product designers as they plan efficient and better usage of the transistors at their disposal. Figure 1.1 shows the scaling of transistor count and operating frequency in ICs [3].

1.1.2   On-Chip Interconnection Schemes

Shared bus was is still the dominant interconnect structure for simple SoC systems. Most buses are bidirectional and devices can send or receive information. The good benefit in bus is that it allows to add new devices easily and facilitates portabilities of peripheral devices between different systems. However, if too many cores are connected to the same bus, the bandwidth of the bus, clock skew and delay can become the bottlenecks.

A new interconnection scheme, known as on-chip network, or NoC, based on *packet switching* approach was proposed [4–6]. NoCs are becoming an attractive option for solving shared bus problems. NoC is a scalable architectural platform with huge potential to handle growing complexity (dozens of cores) and can provide easy reconfigurability, and scalability. The basic idea of NoC is that cores are connected

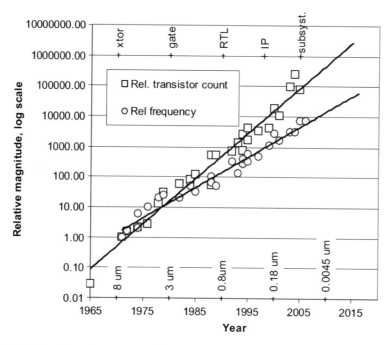

Fig. 1.1 Scaling of transistor counts and operating frequency in ICs. The feature size and design abstraction are also shown in the graph

via a *packet switching* communication on a single chip—similar to the way computers are connected to Internet.

The *packet switching* scheme supports asynchronous transfer of information. Yet, it provides extremely high bandwidth by distributing the propagation delay across multiple switches; thus pipelining the signal transmission. In addition, NoC offers several promising features. First, it transmits packets instead of words. Thus, dedicated address line like in bus systems are not necessary since the destination address of a packet is part of the packet itself. Second, transmission can be conducted in parallel if the network provides more than one transmission channel between a sender and a receiver. Thus, unlike bus-based system, NoC presents theoretical infinite scalability, facilitate IP core reusing, and higher parallelism.

During the last few years, several research groups adopted various concepts from conventional parallel and distributed (Internet) computing world and investigated various design issues related to NoCs. Chapters 4, 5, and 6 will present architecture and design details of such promising interconnects.

1.1.3 Parallelism and Performance

The prevalence of multicore and many-core technologies has brought ubiquitous parallelism and a huge theoretical potential for intensive tasks. Parallelism issue is now affecting all kinds of software development processes. Further, as software, hardware, and applications have evolved, there is a real need to run multiple such tasks simultaneously to benefit from the available hardware capability in multicore and many-core based systems. Thus, a good multicore programming model should be developed and should exploit all types of available parallelism (ILP, DLP, TLP, CLP, etc.) to maximize performance. For example, multimedia applications today often consist of multiple threads or processes. Recall that a thread can be defined as a basic unit of CPU utilization. It consists of a program counter register (PC), CPU state information for the current thread, and other resources such as a Stack (last-in-first-out data structure). However, finding and scheduling parallel instructions or threads is not an easy task since most applications and algorithms are not yet ready to utilize available multicore capabilities.

Most embedded applications are computation-intensive or/and data-intensive types and can only benefit from the full multicore SoC hardware potential if all features on the system level are taken into account. In addition, programmer should exploit different level of parallelisms which are found at several levels in the system. Existing approaches require the programmer/compiler to identify the parallelism in the program and statically create a parallel program using a programming model such as Pthreads (POSIX Threads) [7], MPI (Message Passing Interface) [8], or task programming, expressed in an a high-level language such as C. There are different types of parallelism that a programmer can exploit

- *Bit-Level Parallelism (BLP)*: Bit-Level Parallelism extends the hardware architecture to operate simultaneously on larger data. However, by extending the word length from, for example 8–16, the operation can now be executed by a single operation. This is of course good for performance. Thus, word length has doubled from 4-bit processors through 8, 16, and even 64-bit in advanced processor cores.
- *Instruction-Level Parallelism (ILP)*: ILP is a well known and is (was) an efficient technique for identifying independent instructions and executing them in parallel. Generally, the compiler takes care about finding independent instructions and schedule them for execution by the hardware. Other known techniques are speculative and out-of-order (OoO) execution which are implemented in hardware. Because programs are written in sequential manner, finding independent instructions is not always possible. Some applications, such as for signal processing, can function efficiently and several existing DSP cores can execute eight or even more instructions per cycle and per core (inst/cycle/core).
- *Thread-Level Parallelism (TLP)*: TLP is a software capability that enables a program, often a high-end program to work with multiple threads at the same time instead of having to wait on other threads. TLP can be exploited in single core or also in multicore systems. If used in multicore system, it allows closely coupled cores that share the same memory to run in parallel on shared data structures.

- *Task-Level Parallelism (TaLP)*: TaLP (also known as function parallelism and control parallelism) focuses on distributing execution processes (or threads) across different parallel cores on the same or different data. Most real programs fall somewhere on a continuum between task parallelism and data parallelism. The difficulty with task parallelism is not on how to efficiently distribute the threads, rather is with how to divide the application program into multiple tasks. TaLP approach allows more independent processes to run in parallel, occasionally exchanging messages.
- *Data-Level Parallelism (DLP)*: DLP (also known as loop-level parallelism) allows multiple units to process data concurrently. One such technique implemented in hardware is SIMD (single instruction multiple data). In multiprocessor/multicore system, data parallelism is achieved when each core performs the same task on different pieces of distributed data. Data parallelism is where multicore plays an important role. Performance improvement depends on how many cores are able to work on the data at the same time. For example, consider adding two matrices using two cores (core0 and core1). In a data parallel implementation, core0 could add all elements from the top half of the matrices, while core1 could add all elements from the bottom half of the matrices. Since the two cores work in parallel, the job of performing matrix addition would take one half the time of performing the same operation in serial using one single core.

Since multicore-based systems mainly exploit TLP approach (of course ILP can be also exploited within a single core in a given multicore-based system), we will only focus on this parallelization technique. In order to support TLP, there are several software and hardware approaches that can be used. One approach involves using a preemptive multitasking operating system (OS). This approach involves the use of an interrupt mechanism which suspends the currently executing process and invokes the OS scheduler to determine which process should be executed next. As a result, all processes will get some amount of CPU time at any given time. The OS kernel can also initiate a context switch to satisfy the scheduling policy's priority constraint, thus preempting the active task. The other known approach to address TLP is called Time-slice multi-threading. This approach allows software developers to hide the latency associated with I/Os by interleaving the execution of multiple threads. But the main problem of this approach is that it does not allow for parallel execution, because only one instruction stream can run on a processor at a time.

A more efficient approach for TLP is called simultaneous multi-threading (SMT), or hyper-threading (HT) as called by Intel [9]. The goal of this approach is to efficiently utilize system's resources. SMP makes a single processor appears, from the programmer's view, as multiple logical processor cores. This means, instructions from more than one thread can be executing in any given pipeline stage at a time. This is done without great changes to the main basic building blocks of a processor.

The modern approach for SMP programming is to increase the number of physical processor cores in a computer system or the number of cores in a single die (multicore). As we earlier said, this shift becomes now possible due to the advance of semiconductor technology, which allows the integration of several cores in a single

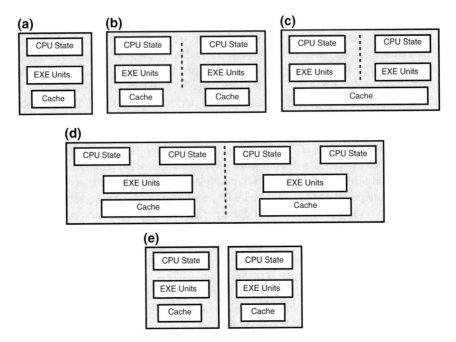

Fig. 1.2 Different ways for exploiting parallelism over various system organization: **a** Single core, **b** Multicore with separate caches, **c** Multicore with shared cache, **d** Multicore with simultaneous threading, **e** Multiprocessor

chip. Integrated cores have their own set of execution and architectural resources and may or may not share a large on-chip cache for better program locality exploitation. For application with large number of threads, individual cores may be implemented with SMP support (see Fig. 1.2d).

1.1.4   Parallel Hardware Architectures

Parallel hardware is becoming an important component in computer processing technology. Recently, there are many dual or quad-core CPUs and graphics processing units (GPUs) on the desktop computer market, and many MCSoC solutions are also in the embedded computing markets. Before we start discussing about multicore SoC architectures, let us first review the different types of parallel hardware architectures, including multiprocessor, dual-core, multicore, SoCs and FPGAs.

Multiprocessors: Multiprocessor systems contain multiple CPU cores that are not on the same chip. These systems were made common in the 1990 s for the purpose of IT servers. Today, multiprocessors are commonly found on the same physical board and connected through a high-speed communication interface.

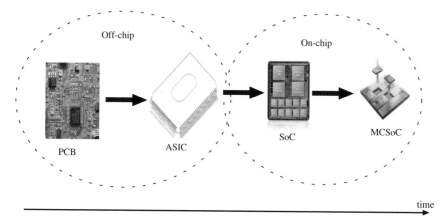

Fig. 1.3 From PCB to MCSoC

Dual-Core and Multicore Processors: Dual-core processors are two CPUs on a single chip (see Fig. 1.2b). Multicore processors are a family of processors that contain any number of multiple CPUs on a single chip, such as 2, 4, and 8. The challenge with multicore processors is in the area of porting existing sequential software (or writing new parallel model) so that it can benefit from the large number of available cores.

FPGAs: A Field Programmable Gate Arrays (FPGAs) is a device that contains a matrix of reconfigurable gate array logic circuitry. When a FPGA is configured, the internal circuitry is connected in a way that creates a hardware implementation of the software application. Unlike ASIC processors, FPGAs use dedicated hardware for processing logic and generally do not have an operating system.

SoC: A system-on-chip (SoC) is an integrated circuit (IC) that integrates all components of a system (generally embedded system) into a single chip. SoC consists of several building blocks including analog, digital, mixed-signal, and radio-frequency functions. These blocks are connected by either a custom or an industry-standard bus such as AMBA [10].

A SoC is quite different from the so called microcontroller. Microcontrollers (i.e., 8051) typically have small RAM memory and are based on low performance processors, whereas an SoC is typically used with more powerful cores, and reconfigurable modules such as a FPGA device. Early SoCs used an interconnect paradigm inspired by the rack-based microprocessor systems of earlier days. Current SoCs use more advanced and scalable interconnects, such as network-on-chip approach (discussed later in Chaps. 4 and 5). Figure 1.3 illustrates the evolution of electronic circuits from simple PCB circuit of earlier days to a state-of-the-art complex multicore SoC system.

1.1.5 The Need for Multicore Computing

As the computing needs of each processor type grew rapidly, the first response of the computer architects and semiconductor companies was to increase the speed of the CPU. Higher performance was mainly achieved by refining manufacturing processes to improve the operating speed. However, this method requires finding solutions for increased leakage power and other problems, making it unable to keep pace with the current rate of evolution or revolution.

After several decades of single-core processor devices production, major CPU makers, such as Intel and AMD, decided to switch to multicore processor chips because it was found that several smaller cores running at a lower frequency can perform the same amount of work without consuming as much energy and power. More precisely, this shift started when Intel's hardware engineers lunched the Pentium 4; at that time, they expected single processor chip to scale up to 10 GHz or even more using advanced process technologies below 90 nm. However, they did not achieve their expectation since the fastest processor never exceeded 4 GHz. As a result, the trends followed by all major hardware makers is to use a higher number of slower cores, building parallel devices made with denser chips that work at low clock speed.

Of course, this revolution could not be achieved without an enormous progress in the semiconductor technologies. That is, the exponential increase in the number of transistors on a die is made possible by the progressive reduction in the characteristic dimensions of the integrating process, from the micrometer resolutions of past decades (with tens of thousands transistors/chip in the 80s, until the recent achievement below hundred manometers (with more than a hundred millions transistor/chip).

Nowadays, semiconductor and hardware companies are fabricating devices realized with technologies down to 45 nm and even less. The merit of reducing dimensions of a chip lies not only in the higher number of gates that can fit on the chip, but also in the higher working frequency at which these devices can be operated. If the distance among every gate becomes small, propagation signals have a lower path to cover, and the transitory time for a state transition decreases, allowing a higher clock speed.

1.1.6 Multicore SoCs Potential Applications

To simplify the discussion, we summarize the potential multicore SoCs applications in Fig. 1.4. The above applications are mainly attractive for embedded systems market. Some of them are also attractive for desktop applications. To meet the requirements of low cost, high performance, and small size, multicore approach plays an important role for system architecture and development. For embedded systems segment, virtually most semiconductor houses are developing systems based on multicore SoC approach. Such multicore SoCs are growing day-after-day and are

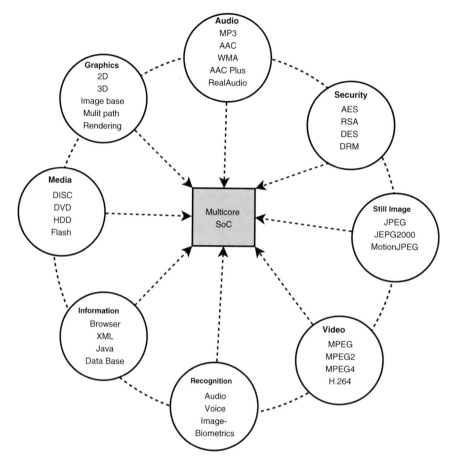

Fig. 1.4 Multicore SoC potential applications

starting to find acceptance in various applications including, real-time mission critical, industrial automation, medical equipment, consumer electronic devices (PDAs,
cellphones, laptops, cameras, etc.) and high-performance computing as shown in
Fig. 1.4.

High-end smartphones already contain a plethora of micro-processors (MPUs)
and digital signal processors (DSPs) to provide advanced modem and application
processing, as well as WiFi, GPS, and Bluetooth functionality.

Another important application is the multimedia domain. Multimedia applications with high-definition audio and video are being provided by embedded systems
such as car navigation, cellular phones, and digital televisions. Multimedia schemes
in general can be partitioned in stream-oriented, block-oriented, and DSP-oriented
functions, which can all run in parallel on different cores. Each core can be used to

run a specific class algorithms, and individual tasks can be mapped efficiently to the appropriate core.

1.2   Multicore SoC Basics

Multicore SoCs are generally constructed with homogeneous or heterogeneous cores. Homogeneous cores are all exactly the same: equivalent frequencies, cache sizes, functions, etc. However, each core in a heterogeneous system may have a different function, frequency, memory model, etc. Homogeneous cores are easier to produce since the same instruction set is used across all cores and each core contains the same hardware. Each core in a heterogeneous multicore SoC, such as the case of CELL processor [11], could have a specific function and run its own specialized instruction set. This model could also have a large centralized core built for generic processing and running a Real-time Operating System (RTOS), a core for graphics, a communications core, an audio core, a cryptography core, etc.

A heterogeneous multicore SoC system is generally more complex to design, but may have better performance, and thermal power benefits that outweigh its complexity. A key difference with classic processor architecture is that the SoC model distinguishes two kinds of processor cores: (1) those used to run the end application, and (2) those dedicated to execute specific functions that could have been designed in hardware. The instruction set architectures (ISAs), programming, and interfacing of these two kinds of processor cores are quite different. Figure 1.5 shows typical multicore SoC architectural view.

1.2.1   Programmability Support

General applications usually consist of several tasks that can be executed on different cores in parallel or concurrently. For example, a multimedia application includes two

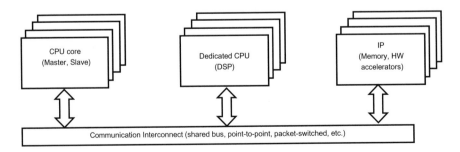

Fig. 1.5 Typical multicore SoC architectural view

concurrent tasks: an audio decoder task and a video decoder task. The task itself may consist of two types of parallelism: (1) functional parallelism and (2) loop-level parallelism. Therefore a multicore platform is needed for such application, and the main design challenge is how to exploit parallelism.

Programmability is also needed for supporting multiple standards and algorithms. For example, some digital video applications require support for multiple video standards, resolutions, and quality. It is easier to implement these on a programmable system. A programmable system can provide the designer the ability to customize a specific algorithm as necessary. This flexibility provides the application's developer with more control of the application.

A multicore SoC may have special instructions to speed up some applications. For example special instructions are implemented on a DSP core to accelerate operations such as: 32-bit multiply instructions (for extended precision computation), expanded arithmetic functions (to support FFT and DCT algorithms), double dot product instructions (for improving throughput of FIR loops), parallel packing instructions, and Enhanced Galois Field Multiply (EGFM).

1.2.1.1 Hardware Accelerators

Hardware accelerator is used on multicore SoCs as a way to efficiently execute some classes of algorithms. There are many applications that have algorithmic functions that do not map very well to a given architecture. Hardware accelerators can be used to solve this problem. Also, a conventional storage model may not be appropriate to execute these algorithms effectively. A specialized hardware accelerator can be built and performs bit manipulation efficiently which sits next to the CPU for bit manipulation operations.

Fast I/O operations are another area where a dedicated accelerator with an attached I/O peripheral will perform better. Finally, applications that are required to process streams of data do not map well to the traditional CPU architecture, especially those that implement caching systems. A specialized hardware accelerator with special fetch logic can be implemented to provide dedicated support to these data streams.

1.2.2 Software Organization

Each kind of multicore SoC employs different software organization. The application software generally consists of several layers on top of the hardware as shown in Fig. 1.6. For software designers, multicore SoC approach presents the interesting challenge of enabling applications to obtain all the processing power available from these multicore environments. How can developers make sure their applications scale linearly with the available cores, as well as fully utilize the other SoC hardware building blocks ? The scalability question is still a real science issue for many applications.

Fig. 1.6 Software layers on
top of the hardware

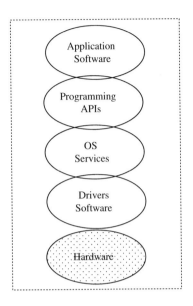

1.2.3 Programming Multicore Systems

Given the various available multicore platforms, choosing the classical programming approach (i.e., OpenMP, MPI) is not always a good decision. The biggest hurdle is the non-deterministic nature of concurrent threads. Thus, to effectively exploit the full power of embedded multicore systems, a more efficient programming model is needed. The standard sequential programming approach cannot be used as it is and should be optimized or extended for such concurrent systems.

Software development for multicore SoCs involves partitioning a given application among the available PEs based on the most efficient computational model. The programmer also has to implement efficient static or dynamic techniques to synchronize between processes. The trend towards multicore systems is motivated by the performance gain compared to single-core systems when a budget on power or temperature or both is given. The performance is expected to further increase with the increasing number of cores if TLP can be fully exploited. The typical goal of threading is to improve the application performance by either increasing the number of work items processed per unit of time (also called throughput) or reducing turnaround time (also called latency).

In order to effectively use *thread* to parallelize a given application, programmer needs a good plan for the overall partitioning of the system and the mapping of the algorithms to the respective processing elements. This may require a lot of trial and error to establish the proper partitioning. There are two main known categories used to do this partitioning: (1) Functional Decomposition—division based on the function of the work, and (2) Data decomposition. In Fig. 1.7, we show two simple examples with functional and data decomposition methods.

Fig. 1.7 Sample OpenMP
code using *section* and
parallel directives:
a Functional decomposition,
b Data decomposition

(a)

```
#pragma omp parallel sections
{
#pragma omp section
    check_scan_attacks();

# pragma omp
    check_denial_service_attacks();

#pragma omp  section
    check_penetration_attacks
}
```

(b)

```
#pragma omp parallel for
{
for (j=0; j<1000 ; j++) {
    process_image(j);
}
```

1.2.4 Multicore Implementations

There are several manufacturers of multicore SoCs. Below, we will describe two well-known multicore systems as an example. Since the target applications of these multicore SoC systems are different, the number of cores, interconnection types, and memory configurations vary widely.

1.2.4.1 CELL Processor

A Sony-Toshiba-IBM partnership built the so-called CELL processor for use in Sony's PlayStation 3 [11]. The CELL system is highly customized for gaming/graphics rendering which means superior processing power for gaming applications. The CELL architecture is a heterogeneous multicore processor that combines a dual-threaded, dual-issue, 64-bit Power-Architecture compliant Power processor element (PPE) with eight newly architected synergistic processor elements (SPEs) an on-chip memory controller, and a controller for a configurable I/O interface [11]. These units are interconnected with a coherent on-chip element interconnect bus (EIB). Extensive support for pervasive functions such as power-on, test, on-chip

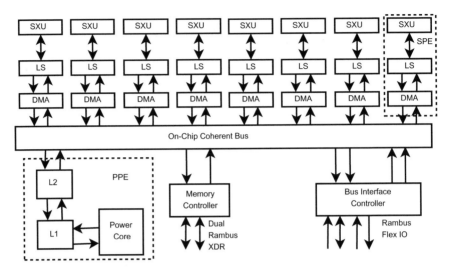

Fig. 1.8 Heterogeneous Multicore CELL Organization

hardware debug, and performance-monitoring functions are also included. With CELL's real-time broadband architecture, 128 concurrent transactions to memory per processor are possible.

In CELL architecture (see Fig. 1.8), Direct Memory Access (DMA) is used to transfer data between local storage and main memory which allows for the high number of concurrent memory transactions. Other interesting features of this architecture are the Power Management Unit (PMU) and Thermal Management Unit (TMU). The PMU allows for power reduction in the form of slowing, pausing, or completely stopping a unit. The TMU consists of one linear sensor and ten digital thermal sensors used to monitor temperature throughout the chip and provide an early warning if temperatures are rising in a certain area of the chip [11].

1.2.4.2   Tilera TILE64

Tilera has developed a multicore chip with 64 homogeneous cores set up in a grid [12, 13]. The family of multicore processors delivers high computing performance and targeted for embedded applications. The processor features 64 identical processor cores (tiles) interconnected with a so-called iMESH on-chip network. Each tile works as a full-featured processor, including integrated L1 and L2 caches and a non-blocking switch that connects the tile into the mesh. Each tile can independently run a full OS, or a group of multiple tiles can run a multi-processing OS such as SMP Linux. An application that is written to take advantage of these additional cores will run far faster than if it were run on a single core.

The TILE64 also includes on-chip memory and I/O controllers. Like the CELL processor, unused tiles (cores) can be put into a sleep mode to further decrease power consumption. The TILE64 uses a three-way VLIW pipeline to deliver 12 times the instructions as a single-issue, single-core processor. When VLIW is combined with the MIMD processors, multiple operating systems can be run simultaneously and advanced multimedia applications such as video conferencing and video-on-demand can be run efficiently [14]

1.3 Multicore SoCs Design Challenges

The introduction of multicore processors signals a major shift in the structure and design ways of all computing platforms. Before this shift, almost all embedded software could be written with the assumption that there is only a single processor core and where multiple processors were involved, they were either relatively loosely coupled or were used in easily parallelized applications.

While multicore systems will change this model somewhat, there is a real expectation that the number of cores will grow rapidly, roughly doubling with each processor generation (Moore's Law still valid). This growth will create unique challenges for run-time systems and compilers. If multiple cores on a processor share a cache, contention for the shared cache memory and cache coherence are major issues.

Power and temperature management are also two concerns that can increase exponentially with the addition of multiple cores. The other issue is the problem of using a multicore processor to its full potential. Applications should be written in a good manner so that different parts of the program runs concurrently. Finally the necessity to move beyond parallel computing paradigm and towards heterogeneous embedded multicore distributed systems will likely drive changes in how embedded software will be created.

1.3.1 Cache Coherence

Allowing multiple processors to share memory complicates the design of the memory hierarchy in a multicore system. Cache coherency, or cache consistency, is a big concern in this multicore environment. Since each core has its own cache, the copy of the data in that cache may not always be the most up-to-date version. For example, imagine a processor with two cores where each core brought a block of memory into its private cache. One core writes a value to a specific location. When the second core attempts to read that value from its cache, it will not have the updated copy unless its cache entry is invalidated and a cache miss occurs. This cache miss forces the second core's cache entry to be updated. This is a real trouble for the correctness of the application being executed.

A system is said to be coherent if all copies of the main memory location in multiple caches remain consistent when the contents of that memory location are modified. A cache coherency protocol (discussed later in Chap. 3) is the mechanism by which the coherency of the caches is maintained. Maintaining coherency means taking special actions when one core writes to a block of data that exists is other caches.

1.3.2   Power and Temperature

While multicore systems may limit power consumption in some areas, they present real challenges to energy management paradigms optimized for single chip systems. In particular, multicore limits the scope and capability of DVFS (dynamic voltage and frequency scaling) because most SoC subsystems share power supplies and clocks. As a result, scaling the operating voltage of one of several SoC subsystems may limit its ability to use local buses to communicate with other subsystems, and to access shared memory. Clock frequency scaling of a single SoC subsystem also presents a big challenges especially for synchronous buses.

To lessen the heat generated by multiple cores on a single chip, the chip is architected so that the number of hot spots does not grow too large and the heat is spread out across the chip. For example, the majority of the heat in the CELL processor is dissipated in the Power Processing Element and the rest is spread across the Synergistic Processing Elements. We will discuss in Chap. 8 in more details power optimization techniques.

1.3.3   Multi-threading and Memory Management

The other important challenge is in using multi-threading or other parallel processing techniques to get the most performance out of the multicore system. Except Java, there are no widely used commercial development languages with multi-threaded extensions [15]. To use multi-threading technique in a given multicore system, programmers have to write applications with subroutines able to be run in different cores, meaning that data dependencies will have to be resolved and applications should be balanced. If one core is being used much more than another, the programmer is not taking full advantage of the multicore system. Microsoft and Apple's newest operating systems can run on up to 4 cores, for example [15, 16].

On multicore SoC system, actual computing is not a problem since there are many processing elements. But, memory bandwidth remains the bottleneck because typical systems use a common bus which is shared by all processor cores. Therefore, efficient memory management is very critical for a scalable application on multicore SoCs.

1.3.4 On-Chip Interconnection Networks

Extra memory will be useless if the amount of time required for memory requests does not improve as well. Currently, on-chip interconnection networks are mostly implemented using buses, where several masters and slaves can be connected to a shared bus. However, an arbiter is needed with a bus to manage multiple requests. A bus arbiter periodically examines accumulated requests from the multiple master interfaces and grants access to a master using arbitration mechanisms specified by the bus protocol. Bus has simple topology, low area, low cost, and easy to build. The disadvantages of shared bus architecture are larger load per data bus line, longer delay for data transfer, large power consumption, and lower bandwidth. To this end, redesigning the interconnection network between cores is a major focus of chip manufacturers.

1.3.5 Reliability Issues

Emerging embedded applications running on multicore SoCs are getting more and more complex, demanding good architectures to ensure sufficient bandwidth for any transaction between memories and cores as well as communication between different cores on the same chip. The significant heterogeneity in multicore SoCs which are likely to mix logic layers with memory layers and even more complex technologies increases the fault's probability in a system. As a result, multicore systems are becoming susceptible to a variety of faults caused by crosstalk, impact of radiations, oxide breakdown, and so on. A simple failure in a single transistor caused by one of these factors may compromise the entire system reliability where the failure can be illustrated in corrupted message delivery, time requirements unsatisfactory, or even sometimes the entire system collapse.

To ensure their correct functionality and reliability, multicore SoCs systems must be fault-tolerant to any short-term malfunction or permanent physical damage to ensure correct functionality while minimizing the performance degradation as much as possible.

1.4 Chapter Summary

Multicore SoCs are architected to adhere to reasonable power consumption, heat dissipation, and cache coherence protocols. In order to use a multicore system at full capacity, the applications must be multi-threaded. However, the difficult task is how to write parallel programs to exploit multicore systems. In addition, the memory systems and interconnection networks also should be carefully designed. This chapter introduced fundamental concepts about multicore SoCs and their design challenges.

References

1. S.R. Vangal et al., An 80-tile sub-100-w teraflops processor in 65-nm CMOS. IEEE J. Solid-State Circuits **43**(1), 29–41 (2008)
2. G. Moore, Cramming more components onto integrated circuits. Electronics Magazine. p. 4. Retrieved 11 Nov. 2006
3. A. Ben Abdallah, *Multicore Systems-on-Chip: Practical Hardware/Software Design*, 2nd edn. (Atlantis, 2013). ISBN-13: 978-9491216916
4. A. Ben Abdallah, M. Sowa, Basic network-on-chip interconnection for future gigascale MCSoCs applications: communication and computation orthogonalization, in *Proceedings of Tunisia-Japan Symposium on Society, Science and Technology (TJASSST)*, 4–9 December 2006
5. W.J. Dally et al., Route packets, not wires: on-chip interconnection networks, in *Proceedings of the DAC*, (2001) pp. 684–689
6. A. Habibi, M. Arjomand, H. Sarbazi-Azad, Multicast-aware mapping algorithm for on-chip networks, *19th International Euromicro Conference on Parallel, Distributed and Network-Based Processing*, (2011), pp. 455–462
7. Pthread Standard: http://standards.ieee.org/findstds/
8. The Message Passing Interface (MPI) Standard, http://www.mcs.anl.gov/research/projects/mpi/
9. D. Koufaty, D.T. Marr, Hyperthreading technology in the netburst microarchitecture, Micro IEEE. **23**(2), 56–65 (2003)
10. ARM, AMBA Overview, http://www.arm.com (2007)
11. B. Flachs et al., The microarchitecture of the streaming processor for a cell processor, in *Proceedings of the IEEE International Solid-State Circuits Symposium*, (2005), pp. 184–185
12. S. Bell, B. Edwards, J. Amann, Tile64-processor: A 64-core soc with mesh interconnect, *Solid-State Circuits*, (2008)
13. Tilera, TILE64 Processor Family, http://www.tilera.com/products/processors.php
14. Tilera, Tile 64 Product Brief, Tilera, (2008)
15. M. Creeger, Multicore CPUs for the masses. QUEUE. (2005)
16. D. Geer, For Programmers, Multicore Chips Mean Multiple Challenges. Computer. (2007)

Chapter 2
Multicore SoCs Design Methods

Abstract The strong demand for low-power and high-performance multicore systems on chip (MCSoCs) requires quick turn around design methodology. Thus, there is a clear need for efficient methodology for the design of these systems on platforms implementing both hardware and software modules. This chapter describes conventional multicore SoC design methods in details. It also describes a so-called scalable core-based methodology for systematic design environment of application-specific heterogeneous multicore SoC architectures. Although the methodology presented here is general and not limited to special architecture, we will consider a real synthesizable core as a case study to make the discussion easy.

2.1 Introduction

Systems-on-chip designs have evolved from fairly simple uni-core, single memory designs to complex multicore SoCs consisting of tens or hundreds of cores in a single chip. As more and more cores are integrated into these chips to share the ever increasing processing load, the main challenges lie in how to efficiently and quickly integrate these cores together into a single system capable of leveraging their individual flexibility. Moreover, for better inter-core communication, the multicore system requires high-performance communication architectures and efficient communication protocols, such as hierarchical bus [1, 2], point-to-point connection [3], time division multiplexed access (TDMA) based bus [4], or packet-switching networks [5].

Recently, SoC design methods tend toward mixed hardware/software codesigns targeting multicore SoCs for specific applications [6–8]. To decide on the lowest cost mix of cores, designers must iteratively map the device's functionality to a particular hardware/software partition and target architecture (platform). When a designer wants to explore different system architectures, the interfaces must be redesigned. This method may lead to a narrow application domain. In addition, managing all these details is time-consuming that designers typically cannot afford to evaluate several different implementations.

Automating the interface generation is an alternative solution and a critical part of the development of embedded system' synthesis tools. Most existing automation

© Springer Nature Singapore Pte Ltd. 2017
A. Ben Abdallah, *Advanced Multicore Systems-On-Chip*,
DOI 10.1007/978-981-10-6092-2_2

algorithms implement the system based on a standard bus protocol (input/output interface) or based on a standard component (processing) protocol. Recent works have used a more generalize model consisting of heterogeneous multicore with arbitrary communication links. The SOS algorithm [9] uses an integer linear programming approach. The co-synthesis algorithm, developed in [10], can handle multiple objectives such as cost, performance, power, and fault tolerance. Such design methods allow only limited automation and designers resort to manual architecture design which is time-consuming and error-prone.

There are two fundamental steps needed for MCSoC design: (1) selection and construction of a target multicore platform, known as design space exploration phase, and (2) development of the parallel software for exploiting the application parallelism on the selected platform, known as parallel software development phase. We will describe these hardware and software design phases in the following two sections.

2.2 Design Space Exploration

There are various design axes that define the design space of multicore platforms, which include processor architectures and numbers, memory configuration, communication architectures, hardware accelerators, and so on. To determine the target platform, we need a technique that quickly evaluates the expected performance of each candidate and explores the wide design space without actual hardware implementation. Further, the gate densities achieved in current ASIC and FPGA devices give designers enough logic elements to implement all functionalities on the same chip by mixing self-design modules with third party ones [4, 7, 11]. This possibility opens new horizons especially for embedded systems where space constraints are as important as performance. The most fundamental characteristic of a SoC is complexity. The SoC is generally tailored to the application rather than general-purpose chip, and may contain memory, one or several specialized cores, buses, and several other digital functions. Therefore, embedded applications cannot use general-purpose computers (GPPs) either because a GPP machine is not cost effective or because it cannot provide the necessary requirements and performance. In addition, a GPP machine cannot provide reliable real-time performance.

In Fig. 2.1, a typical multicore SoC architecture block diagram is shown. This typical model is made of a set of cores communicating through an AMBA communication architecture [1]. The communication architecture constitutes the hardware links that support the communication between cores. It also provides the system with the required support for the general data transfer with external devices common to most applications. Inter-component link is often in the critical path of such a system and is a very common source of performance bottlenecks [12]. Thus, it becomes imperative for system designers to focus on exploring the communication design space.

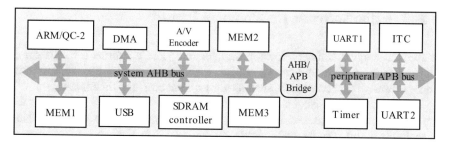

Fig. 2.1 SoC typical architecture

Conventional SoC architectures are generally classified into two types: single-core-based and multicore-based systems. Single-core architecture consists of a single CPU core and one or several ASICs. A master–slave synchronization pattern is adopted in this type. The single-core SoC type can only offer a restricted performance capability in many applications because of the lack of true parallelism.

A multicore SoC architecture is a system that contains multiple CPU cores and also one or several ASICs. In term of performance, multicore SoCs perform better for several embedded applications. However, these systems generally introduce new challenges: first, the inter-processor communication may require more sophisticated networks than a simple shared bus, and second, the architecture may include more than one master processor. In both types, high processing performance is required because most of the applications for which SoCs are used have precise performance requirements deadlines; this is different from conventional general-purpose computing.

In general, the architectures used in conventional methods of multicore SoC design and custom multicore architectures are not flexible enough to meet the requirements of different application domains (e.g., only point-to-point or shared bus communication is supported) and not scalable enough to meet different computation needs and different complexity of various applications. A promising approach was proposed in [10]. This method is a core-based solution, which enables integration of heterogeneous processors and communications protocols by using abstract interconnections. Behavior and communication must be separated in the system specification. Hence, system communication can be described at a higher level and refined independently of the behavior of the system. There are two known component-based design approaches: (1) usage of a standard bus (i.e., IBM CoreConnect) protocol, and (2) usage of a standard component protocol [6–8]. For the first approach, a wrapper is designed to adapt the protocol of each component to CoreConnect protocol. For the second case, the designer can choose a bus protocol and then design wrappers to interconnect components using the above protocol.

2.3 Parallel Software Development Phase

Embedded parallel software development for multicore platforms involves parallel programming for homogeneous and heterogeneous multicore SoC architectures under several design constraints such as power, area, cost, and timeliness.

The sequential Von Neumann programming model is not a good option for the multicore-based systems because it simply cannot exploit the huge parallelism which is available in different forms in multicore platforms. Thus, it is clear that we now need new programming models and corresponding software development tools that are capable of exploiting all forms of available parallelism. Recently, big efforts have been made to develop methods and tools that solve the design problems of multicore SoCs targeted for various applications and under several design constraints. Below, we will describe these methods in details.

2.3.1 Compiler-Based Schemes

In compiler-based schemes, the sequential Von Neumann program is used as input, where all specifications are defined (Phase 1). Then, a parallelizing compiler automatically parallelizes (Phase 2) the source code (or binary) as illustrated in Fig. 2.2. Using several parallelizing techniques, this phase (Phase 2) analyzes the input code and finds parallel regions. More specifically, Phase 2 parallelizes the serial code. A well-known technique is to identify all loops and examine their dependencies by analyzing indexes. The mapper, then, transforms each parallel region into a set of concurrent tasks and maps them onto multiple cores (Phase 3).

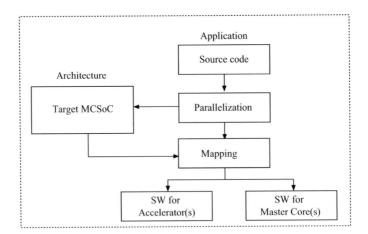

Fig. 2.2 Compiler-based scheme

2.3.2   *Language Extensions Schemes*

The *language extension schemes* require that application programmer provides all parallelism information as well as where and how to parallelize the code with language extension that has annotations and/or additional application programming interfaces (APIs). As a result, compilers in *language extension schemes* can focus on exploiting the specified parallelism according to the the target platform.

2.3.2.1   Language Extension with Annotations

The main merit of the language extension with annotations approach is simplicity. That is, it simplifies the compiler's job by relieving the burden of extracting parallelism while it gives only a little overhead of annotations to the software developer.

The Open Multiprocessing Standard [13] is an example of language extension with annotations. OpenMP is a widely used API for parallel programming and is attractive because programmers can continue using their familiar programming model while reusing their existing codes.

As an example, suppose a programmer is writing a ray tracing program, which goes through each pixel of the screen, and using lighting, texture, and geometry information, the color of that pixel is determined. The program goes on to the next pixel and repeats (loops) the process. The calculation for each pixel is completely separate from the calculation of any other pixel, therefore making this program highly suitable for OpenMP. The code for the above example is shown in Fig. 2.3. This piece of code simply goes through each pixel of the screen, and calls a function, RenderPixel, to determine the final color of that pixel. Note that the results are simply stored in an array. Because each pixel is independent of all other pixels, and because RenderPixel is expected to take a noticeable amount of time, this small snippet of code is a prime candidate for parallelization and can be simply annotated with OpenMP directive: #*pragma omp parallel for*.

We have to note here that OpenMP standard was originally developed for symmetric multiprocessor (SMP) computers with shared memory. Recently, it was ported to heterogeneous multicore platforms, such as in IBM Cell processor [14]. GNU GCC also adopted the GOMP OpenMP implementation. Thus, many GCC-enabled multicore processors now support OpenMP [15]. The cell processor is a heterogeneous multicore processor with one Power Processing Engine (PPE) core and eight Synergistic Processing Engine (SPE) cores. Each SPE has a directly accessible small local memory (256K), and it can access the system memory through DMA operations. Programming cell system is difficult since an SPE core has a small local memory and accesses the system memory only through DMA operations. The other difficulty comes from the availability of several layers of parallelism in the architecture, including heterogeneous cores, multiple SPE cores, multi-threading. Cell compiler is built upon an IBM XL compiler therefore translates the parallel region into a set of concurrent tasks that run on the SPE cores with a control task that schedules the SPE tasks [14].

Fig. 2.3 Parallel for loop
with OpenMP

```
for(int x=0; x < width; x++)
{
  for(int y=0; y < height; y++)
  {
    finalImage[x][y] = RenderPixel(x,y, &sceneData);
  }
}
                      (a) Before parallelization

#pragma omp parallel for
for(int x=0; x < width; x++)
{
  for(int y=0; y < height; y++)
  {
    finalImage[x][y] = RenderPixel(x,y, &sceneData);
  }
}
                      (b) After parallelization
```

2.3.3 Language Extensions with APIs

In this scheme, a software developer writes a parallel program with specifically
defined APIs for parallel execution. Compared with the annotation scheme, the
APIs-based approach allows more low-level control of parallelism by the software
developer. Although this scheme has better performance, it requires that the programmer manually discovers the parallel regions, distributes the code and data to the
processors, and restructures the code using the APIs.

Message passing interface [16] is an example of the language extension with APIs
since it started to find its use in embedded heterogeneous multicore SoCs.

2.3.4 Model-Based Schemes

Model-based schemes are advocated for multicore and MCSoC design since they
simplify the application behavior and reveals the top-level structure of the behavior; this eliminates the complex low-level implementation details. In this scheme,
the software developer determines which model of computation is used to capture
application algorithms. For example, the actor based models are used to specify
the computation-oriented applications and the FSM (finite state machine) model for
control-oriented applications.

2.4 Generic Architecture Template (GAT) for Real Multicore SoC Design

In this section, we will describe a design method based on a so-called generic architecture template (GAT), where both processing and input/output interface may be customized to fit the specific needs of the application. GAT design method enables a designer to make a basic architecture design without detailed knowledge of the architecture.

A high-performance synthesizable soft-core architecture, called QueueCore, is also presented here and is used as a task-distributor-core (TDC) in the a multicore SoC system design. The system may consist, then, of multiple processing cores of various types (i.e., QueueCore(s), general-purpose processor(s), domain specific DSPs, and custom hardware), and communication links. The ultimate goal of the above systematic design automation and architecture generation is to improve performance and the design efficiency of large-scale heterogeneous multicore SoC.

2.4.1 Target Multicore SoC Platform

The target model of the architecture consists of CPUs (i.e., QueueCore (QC-2), GPPs), hardware blocks, memories, and communication interfaces. The addition of new core will not change the main principle of the proposed methodology. The core is connected to the shared communication architecture via communication network, which maybe of whatever complexity from a single bus to a network with complex protocols. However, to ensure modularity, standard and specific interfaces to link cores to the communication architecture should be used. This gives the possibility to design separately each part of the application. Reader can refer to [17] for more details about a modular design methodology. One important feature of the above method is that the generic assembling scheme largely increases the architecture modularity. Figure 2.4 shows a typical instance of the platform made of four cores (2*QC-2 cores and 2*SH cores). The QC-2 core is a special purpose synthesizable core (described in details in Sect. 2.4.3).

The designer can configure: the number of CPUs, I/O ports for each processor and interconnections between cores, the communication protocol and the external peripherals. The communication interface depends on the core attributes and on the application-specific parameters. The communication interface connects a given core to the communication architecture and consists of two parts: the first part specific to the core's bus and the second part is generic and depends on communication protocols and on the number of communication channels used. This structure allows the *isolation* of the cores from the communication network.

Each interface module acts as a coprocessor for the corresponding core. The application dependent part may include several communication channels. The arbitration is done by the CPU-dependent part and the overhead induced by this communication

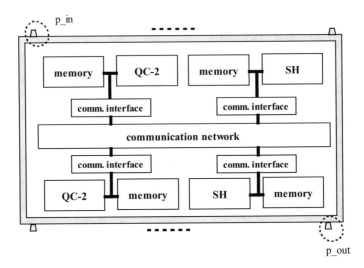

Fig. 2.4 Multicore SoC system platform. This is a typical instance of the architecture, where the addition of a new core will not change the principle of the methodology

coprocessor depends on the design of the basic components and may be very low. The use of this architecture for interfaces provides huge flexibility and allows for modularity and scalability.

2.4.2 Design Method

In this methodology, the application-specific parameters should be used to configure the architecture platform and an application-specific architecture is produced. These parameters are determined from an analysis of the application to be designed. The design flow graph (DFG) is divided into 14 *linked tasks* as shown in Fig. 2.5a, b and summarized in Table 2.1. The first task (node T1) defines the architecture platform using all fixed architectural parameters: (1) Network type, (2) Memory architecture, (3) CPU types, and (4) other HW modules. Using the application system level description (second task) and the architectural fixed parameters, the selection of the actual design parameters (number of CPUs, the memory sizes for each core, I/O ports for each core and interconnections, between cores, the communication protocols and the external peripherals) is performed in task 3 (node T3). The outputs of task 3 are: an abstract architecture description (node T7) and a mapping table (node T6). Node T7 is the internal structure of the target system architecture. It contains all the application-specific parameters. The mapping table (T7) contains the addresses allocation and memory map for each core. The complete architecture design task (T8) is

Fig. 2.5 Linked task design flow graph (DFG).
a Hardware related tasks,
b Application related tasks

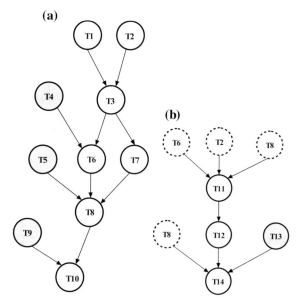

Table 2.1 Linked task description

Task	Description
T1	Define architecture platform
T2	Describe application system level
T3	Select design parameters
T4	Instantiate Pr. att.
T5	Instantiate communication
T6	Mapping table
T7	Describe abstract architecture
T8	Design architecture
T9	Inst. IP cores (Pr.and Mem)
T10	H-SoC synthesis
T11	Software adaptation
T12	Binary code
T13	Pr. and memory emulators
T14	H-SoC validation

linked to the abstract architecture and the mapping table nodes (tasks). Finally, binary programs that will run on the target processors are produced in task 11 (node T11). For validation, cycle accurate simulation for CPUs and HDL (Verilog or VHDL) modeling for other cores/modules can be used for the whole architecture.

2.4.3   QueueCore Architecture

The key idea of the produced order queue computation model is the operands and results manipulation schemes [18, 19]. The queue computing scheme stores intermediate results into a circular queue register (QREG).

A given instruction implicitly reads its first operand from the head of the QREG, its second operand from a location explicitly addressed with an offset from the first operand location. The computed result is finally written into the QREG at a position pointed by a queue tail pointer (QT). An important feature of this scheme is that write-after-read false data dependency does not occur [17, 20–24]. Furthermore, since there is no explicit referencing to the QREG, it is easy to add extra storage locations to the QREG when needed. The other feature of this computing model is its important affect on the instruction issue hardware.

The QC-1 core [18] exploits ILP without considerable effort for heavy run time data dependence analysis, resulting in a simple hardware organization when compared with conventional super scalar processors. This also allows the inclusion of a large number of functional units into a single chip, increasing parallelism exploitation. Since the operands and result addresses of a given static instruction (compiler generated) are implicitly *computed* during run time, an efficient and fast hardware mechanism is needed for parallel execution of instructions. The queue processor implements a so-named queue computation mechanism that calculates operands and result addresses for each instruction (discussed later). The QC-2 core implements all hardware features found in QC-1 core and also supports single precision floating-point accelerator.

2.4.3.1   Hardware Pipeline Structure

The QC-2 supports a subset of the produced order queue processor instruction set architecture [18]. All instructions are 16-bit wide, allowing simple instructions fetch and decode stages and facilitate instructions pipelining. The pipeline's regular structure allows instructions fetching, data memory references, and instruction execution to proceed in parallel. Data dependencies between instructions are automatically handled by hardware interlocks. Below, we describe the salient characteristics of the QueueCore architecture.

(1) *Fetch (FU)*: The instruction pipeline begins with the fetch stage, which delivers four instructions to the decode unit each cycle. This is the same bandwidth as the maximum execution rate of the functional units. At the beginning of each cycle, assuming no pipeline stalls or memory wait states occur, the address pointer hardware of the fetched instructions issues a new address to the data/instruction memory system. This address is either the previous address plus 8 bytes or the target address of the currently executing flow-control instruction.

(2) *Decode (DU)*: The QC-2 decodes four instructions in parallel during the second phase and writes them into the decode buffer. This stage also calculates the number

of consumed (CNBR) and produced (PNBR) data for each instruction. The CNBR and PNBR are used by the next pipeline stage to calculate source and destination locations for each instruction. Decoding stops if a queue becomes full.

(3) *Queue computation (QCU)*: The QCU calculates the first operand (*source*1) and destination addresses for each instruction. The QCU unit keeps track on the current value of the QH and QT pointers. Four instructions arrive to the QCU unit each cycle. To execute instructions in parallel, the QC-2 core must calculate the operands addresses (*source*1, *source*2 and *destination*) for each instruction. Figure 2.6 illustrates QC-2's next QH and QT pointers calculation mechanism. To calculate the *source*1 address, the consumed operands (CNBR) field (port field) is added to the current QH value (QH0). The second operand address in calculated as shown in Fig. 2.7. Similar mechanism is used for the other three instructions. Because the next QH and QT values are dependent on the current QH and QT values, the calculation is performed sequentially. Each QREG entry is written exactly once and it is busy until it is written. If a subsequent instruction needs its value, that instructions must wait until it is written. After QREG entry is written, it is ready.

(4) *Barrier:* The major goal of this unit/stage is to insert barrier flags for all barrier type instructions.

(5) *Issue:* Four instructions are issued for execution each cycle. In this stage, the second operand (*source*2) of a given instruction is first calculated by adding the address *source*1 to the displacement that comes with the instruction. The second

Fig. 2.6 Next QH and QT
pointers calculation
mechanism

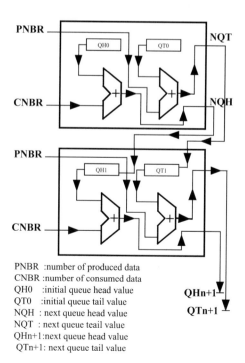

PNBR :number of produced data
CNBR :number of consumed data
QH0   :initial queue head value
QT0   :initial queue tail value
NQH : next queue head value
NQT : next queue teail value
QHn+1:next queue head value
QTn+1: next queue tail value

Fig. 2.7 QC-2's source 2
address calculation

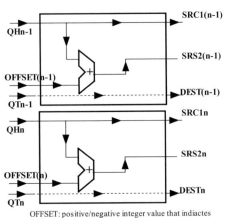

OFFSET: positive/negative integer value that indiactes
the location of SRC2(n-1) from the QH(n-1)
QTn   : queue tail value of instruction n
DESTn : destination location of instruction n
SRC1(n-1): source data 1 of instruction (n-1)
SRC2(n-1): source data 2 of instruction (n-1)

operand's address calculation could be earlier calculated in the QCU stage. However, for a balanced pipeline consideration, the *source2* is calculated in this stage.

An instruction is ready to be issued if its data operands and its corresponding functional unit are available. The processor reads the operands from the QREG in the second half of stage 5 and execution begins in stage 6.

(6) *Execution (EXE)*: The macro-data flow execution core consists of 1 integer ALU unit, 1 floating-point accelerator unit, 1 branch unit, 1 multiply unit, 4 set units, and 2 load/store units.

The load and store units share a 16-entry address window (AW), while the integer unit and the branch unit share a 16-entry integer window (IW). The FPA has its own 16-entries floating-point window (FW). The load/store units have their own address generation logic. Stores are executed to memory in-order.

2.4.3.2 Floating-Point Organization

The QC-2 floating-point accelerator (FPA) is a pipelined structure and implements a subset of the IEEE-754 single-precision floating-point standard [25, 26]. The FPA consists of a floating-point ALU (FALU), floating-point multiplier (FMUL), and floating-point divider (FDIV). The FALU, FMUL, FDIV, and the floating-point queue register (FQREG) employ 32-wide data paths. Most FPA operations are completed within three execution cycles. The FPA's execution pipelines are simple in design for high speeds that the QC-2 core requires. All frequently used operations are directly implemented in the hardware. The FPA unit supports the four rounding modes specified in the IEEE 754 floating-point standard: round toward-to-nearest-even, round toward positive infinity, round toward negative infinity, and round toward zero.

Floating-point ALU implementation: The FALU does floating-point addition, subtraction, compare and conversion operations. Its first stage subtracts the operands exponents (for comparison), selects the larger operand, and aligns the smaller mantissa. The second stage adds or subtracts the mantissas depending on the operation and the signs of the operands. The result of this operation may overflow by a maximum of 1-bit position. Logic embedded in the mantissa adder is used to detect this case, allowing 1-bit normalization of the result on the fly. The exponent data path computes $(E + 1)$. If the 1-bit overflow occurred, $(E + 1)$ is chosen as the exponent of stage 3; otherwise, E is chosen. The third stage performs either rounding or normalization because these operations are not required at the same time. This may also result in a 1-bit overflow. Mantissa and exponent corrections, if needed, are implemented exactly in this stage, using instantiations of the mantissa adder and exponent blocks.

The area efficient FADD hardware is shown in Fig. 2.8. The exponents of the two inputs (Exponent A and Exponent B) are fed into the exponent comparator, which is implemented with a subtractor and a multiplexer. In the pre-shifter, a new mantissa in created by right shifting the mantissa corresponding to the smaller exponent by the difference of the exponents so that the resulting two mantissas are aligned and can be added. The size of the pre-shifter is about $m * log(m) LUTs$, where m is the bit-width of the mantissa. If the mantissa adder generates a carry output, the resulting mantissa is shifted one bit to the right and the exponent is increased by one. The normalizer transforms the mantissa and exponent into normalized format. It first uses a leading-one detector (LD) circuit to locate the position of the most significant one in the mantissa. Based on the position of the LD, the resulting mantissa is left shifted by an amount subsequently deducted from the exponent. If there is an exponent overflow (during normalization), the result is saturated in the direction of overflow and the

Fig. 2.8 QC-2's FADD hardware

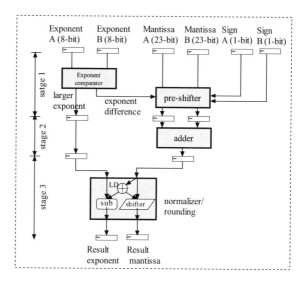

overflow flag is set. Underflows are handled by setting the result to zero and setting an underflow flag.

We have to notice that the LD anticipator can be also predicted directly from the input to the adder. This determination of the leading digit position is performed in parallel with the addition step so as to enable the normalization shift to start as soon as the addition completes. This scheme requires more area than a standard adder, but exhibits reduced latency. For hardware simplicity and logic limitation, our FPA hardware does not support earlier LD prediction.

Floating-point multiplier implementation: The data path of the FMUL hardware is shown in Fig. 2.9. As with other conventional architectures, QC-2's FMUL operation is much like integer multiplication. Because floating-point numbers are stored in sign magnitude form, the multiplier needs only to deal with unsigned integer numbers and normalization. Similar to the FALU, the FMUL unit is a three stages pipeline that produces a result on every clock cycle. The bottleneck of this unit was the $24 * 24$ integer multiplications.

The first stage of the floating-point multiplier is the same denormalization module used in addition to insert the implied 1 to the mantissa of the operands. In the second stage, the mantissas are multiplied and the exponents are added. The output of the module is registered. In the third stage, the result is normalized or rounded.

The multiplication hardware implements the radix-8 modified Booth [27] algorithm. Recoding in a higher radix was necessary to speed up the standard Booth multiplications algorithm since greater numbers of bits are inspected and eliminated during each cycle, effectively reduces the total number of cycles necessary to obtain the product. In addition, the radix-8 version was implemented instead of the radix-4 version because it reduces the multiply array in stage 2.

Fig. 2.9 QC-2's FMUL hardware

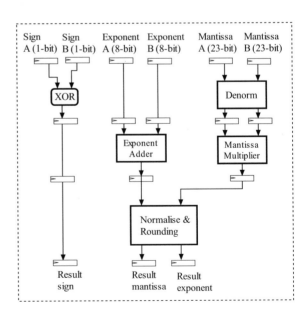

2.4.4 Performance Analysis

In order to estimate the impact of the description style on the target FPGAs efficiency, logic synthesis for FPGAs is explored. The idea of this experiment was to optimize critical design parts for speed or resource optimizations.

Optimizing the HDL description to exploit the strengths of the target technology is of paramount importance to achieve an efficient implementation. This is particularly true for FPGAs targets, where a fixed amount of each resource is available and choosing the appropriate description style can have a high impact on the final resources efficiently [28, 29]. For typical FPGAs features, choosing the right implementation style can cause a difference in resource utilization of more than an order of magnitude [30, 31]. Synthesis efficiency is influenced significantly by the match of resource implied by the HDL and resources present in a particular FPGAs architecture. When an HDL description implies resources not found in a given FPGAs architecture, those elements have to be emulated using other resources at significant cost. Such emulation can be performed automatically by EDA tools in some cases, but may require changes in the HDL description in the worst case, counteracting aim of a common HDL source code base. In this work, our experiments and the results described are based on the Altera Stratix architecture [32]. We selected Stratix FPGAs device because it has a good trade-offs between routability and logic capacity. In addition it has an internal embedded memory that eliminates the need for external memory module and offers up to 10 Mbits of embedded memory through the TriMatrix TM memory feature. We also used Altera Quartus II professional edition for simulation, placement, and routing. Simulations were also performed with Cadence Verilog-XL tool [33].

Figure 2.10 compares two different target implantations for 256 × 33 QREG for various optimizations. Depending on the target implementations device, either logic elements (LEs) or total combinational functions (TCF) are generated as storage ele-

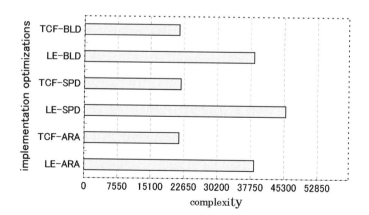

Fig. 2.10 Resource usage and timing for 256*33 bit QREG unit for different coding and optimization strategies

Table 2.2 QC-2 processor design results: modules complexity as LE (logic elements) and TCF (total combinational functions) when synthesized for FPGA (with Stratix device) and Structured ASIC (HardCopy II) families

Descriptions	Modules	LE	TCF
Instruction fetch unit	IF	633	414
Instruction decode unit	ID	2573	1564
Queue compute unit	QCU	1949	1304
Barrier queue unit	BQU	9450	4348
Issue unit	IS	15,476	7065
Execution unit	EXE	7868	3241
Queue registers unit	QREG	35,541	21,190
Memory access	MEM	4158	3436
Control unit	CTR	171	152
Queue processor core	QC-2	77,819	42,714

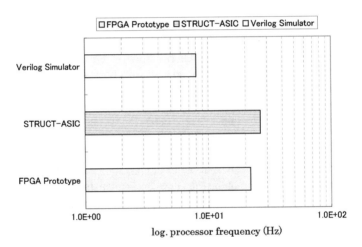

Fig. 2.11 Achievable frequency is the instruction throughput for hardware implementations of the QC-2 processor. Simulation speeds have been converted to a nominal frequency rating to facilitate comparison

ments. Implementations based on HardCopy device, which generates TCF functions give almost similar complexity for the three used optimizations—area (ARA), speed (SPD), and balanced (BLD). For FPGA implementation, the complexity for SPD optimization is about 17 and 18% higher than that for ARA and BLD optimizations, respectively. Table 2.2 summarizes the synthesis results of the QC-2 for the Stratix FPGA and HardCopy targets. The complexity of each core module as well as the whole QC-2 core are given as the number of logic elements (LEs) for the Stratix FPGA device and as the TCF cell count for the HardCopy device (Structured ASIC). The design was optimized for BLD optimization guided by a properly implemented constraint table. We also found that the processor consumes about 80.4% of the total logical elements of the target device.

The achievable throughput of the 32-bit QC-2 core on different execution platforms is shown in Fig. 2.11. For the hardware platforms, we show the processor frequency. For comparison purposes, the Verilog HDL simulator performance has been converted to an artificial frequency rating by dividing the simulator throughput by a cycle count of 1 CPI. This chart shows the benefits which can be derived from direct hardware execution using a prototype when compared to processor simulation. The data used for this simulation are based on event-driven functional Verilog HDL simulation.

2.5   Chapter Summary

SoC designs have evolved from fairly simple single-core designs to complex multi-core SoCs consisting of hundreds of PEs in a single chip. As more and more cores are integrated into these chips, the main challenges lie in how to efficiently and quickly integrate these cores together into a single system capable of leveraging their individual flexibility.

There are two fundamental issues for MCSoC design: (1) design space exploration, and (2) parallel software development. This chapter focused on these two schemes. The chapter also presented a scalable core-based methodology for generic architecture model and a synthesizable 32-bit soft core suitable for high-performance multicore SoC architectures. The presented GAT method should permit a systematic generation of multicore architecture for embedded multicore SoCs.

References

1. K. Diefendorff, K. Dubey, How multimedia workloads will change processor design. IEEE Comput. **30**(9), 43–45 (1997)
2. Y. Liu, S. Chakraborty, W.T. Ooi, A. Gupta, S. Mohan, Workload characterization and cost-quality tradeoffs in MPEG-4 decoding on resource-constrained devices, in *Workshop on Embedded Systems for Real-Time Multimedia* (2005), pp. 129–134
3. M. Loghi, F. Angiolini, D. Bertozzi, L. Benini, R. Zafalon, Analyzing on-chip communication in a mpsoc environment. Proceedings of the Conference on Design, Design Automation and Test in Europe **2**, 16–20 (2004)
4. D. Kulkarani, W.A. Najjar, R. Rinker, F.J. Kurdahi, *Fast area estimation to support compiler optimization in FPGA-based reconfigurable systems, in IEEE Symposium on Field-Programmable Custom Computing Machines* (California, Napa, 2002)
5. A. Ben Abdallah, M. Sowa, Basic network-on-chip interconnection for future gigascale mcsocs applications: communication and computation orthogonalization, *in Proceedings of Tunisia-Japan Symposium on Society, Science and Technology (TJASSST)*, 4–9 Dec 2006
6. R. Ernst, J. Henkel, T. Benner, Hardware-software co synthesis for microcontrollers. IEEE Des. Test 64–75 (1993)

7. A. Jerraya, *Multiprocessor System-on-Chip*, (Morgan Kaufman Publishers, 2005) ISBN:0-12385-251-X
8. C.K. Lennard, P. Schaumont, G. de Jong, A. Haverinen, P. Hardee, Standards for system-level design: practical reality or solution in search of a question?, in *Proceedings of the Design Automation and Test in Europe*, (2000), pp. 576–585
9. S. Prakash, A. Parker, SoS: Synthesis of application-specific heterogeneous multiprocessor systems. J. Parellel Distrib. Comput. **16**, 338–351 (1992)
10. B. Dave, G. Lakshminarayama, N. Jha, COSFA: Hardware-software co-synthesis of heterogeneous distributed embedded system architectures for low overhead fault tolerance, *in Proceedings IEEE Fault-Tolerant Computing Symposium*, (1997), pp. 339–348
11. M. Sheliga, E.H. Sha, Hardware/software co-design with the hms framework. J. VLSI Signal Process. Systems **13**(1), 37–56 (1996)
12. S. Pasricha, N. Dutt, M. Ben-Romdhane, *Constraint-driven bus matrix synthesis for mpsoc, Asia and South Pacific Design Automation Conference (ASPDAC 2006)* (Japan, Yokohama, 2006), pp. 30–35
13. OpenMP: API Specification for Parallel Programming: http://openmp.org
14. K. Obrien, Z. Sura, T. Chen, T. Zhang, Supporting OpenMP on cell. J. Parallel Program. **36**(3), 289–311 (2008)
15. GOMP: An OpenMP Implementation for GCC, Available: http://gcc.gnu.org/projects/gomp
16. The Message Passing Interface (MPI) Standard: http://www.mcs.anl.gov/research/projects/mpi/
17. A. Ben Abdallah, S. Kawata, T. Yoshinaga, M. Sowa, Modular design structure and high-level prototyping for novel embedded processor core, *Proceedings of the 2005 IFIP International Conference on Embedded And Ubiquitous Computing (EUC'2005)*, (Nagasaki, Japan, Dec. 6–9, 2005), pp. 340–349
18. A. Ben Abdallah, M. Arsenji, S. Shigeta, T. Yoshinaga, M. Sowa, Queue processor for novel queue computing paradigm based on produced order scheme, *in Proceedings of HPC, IEEE CS*, July 2004, pp. 169–177
19. A. Ben Abdallah, A. Canedo, T. Yoshinaga, M. Sowa, The QC-2 parallel queue processor architecture. J. Parallel Distrib. Comput. **68**(2), 235–245 (2008)
20. A. Ben Abdallah, M. Masuda, A. Canedo, K. Kuroda, Natural instruction level parallelism-aware compiler for high-performance queuecore processor architecture. J. Supercomput. **57**(3), 314–338 (2011)
21. A. Canedo, A.B. Abdallah, M. Sowa, compiler support for code size reduction using a queue-based processor. Transactions on High-Performance Embedded Architectures and Compilers **2**(4), 269–285 (2009)
22. A. Canedo, A.B. Abdallah, M. Sowa, Compiling for reduced bit-width queue processors. J. Signal Process. Syst. **59**(1), 45–55 (2010)
23. A. Canedo, A.B. Abdallah, M. Sowa, Efficient compilation for queue size-constrained queue processors. J. Parallel Comput. **35**, 213–225 (2009)
24. A. Canedo, A.B. Abdallah, M. Sowa, Design and implementation of a queue compiler. J. Microprocess. Microsyst. **33**(2), 29–138 (2009)
25. IEEE standard for binary floating-point arithmetic, ANSI/IEEE standard 754, (1985)
26. IEEE task P754, A proposed standard for binary floating-point arithmetic, IEEE Comp. **14**(12), pp. 51–62, (1981)
27. A.D. Booth, A signed binary multiplication technique. Quart. J. Mech. Appl. Math. **4**, 23–40 (1951)
28. G. De Micheli, R. Ernst, W. Wolf, *Readings in Hardware/Software co-design*, (Morka Kaufmann Publishers, ISBN: 1-55860-702-1, 2001)
29. D. Gohringer, M. Hubner, V. Schatz, J. Becker, Runtime adaptive multi-processor system-on-chip: RAMPSoC, *in International Symposium on Parallel and Distributed Processing*, 1–7 April 2008
30. A. Alsolaim, J. Becker, M. Glesner, J. Starzyk, Architecture and application of a dynamically reconfigurable hardware array for future mobile communication systems, *in IEEE International Conference on Field-Programmable Custom Computing Machines*, (2000), pp. 205–214

31. Xilinx, Virtex-5 Family Overview, (February 2009)
32. Altera Design Software, http://www.altera.com/
33. Cadence Design Systems, http://www.cadence.com/

Chapter 3
Multicore SoC Organization

Abstract Increasing processing power demand for new embedded consumer applications such as mobile multimedia devices, cell phones, and high definition televisions made convectional single-core SoC-based designs no longer suitable to satisfy high performance and low power consumption demands. Moreover, continuous advancements in semiconductor technology enable us to design more complex multicore systems-on-chip (MCSoCs) composed of tens or even hundreds of IP cores. General purpose CPUs, ASICs, DSPs, memory blocks, and I/O and networking devices on a single MCSoC chip are now possible and necessary for current and future complex applications. Understanding the software and hardware building blocks and the computation power of individual components in these complex MCSoCs are necessary for designing power, performance, and cost-efficient systems. This chapter describes in details the architectures and functions of the main building blocks that are used to build such complex MCSoCs.

3.1 Introduction

With increasing processing power demands of embedded applications and technology advances, MCSoCs become prevalent in embedded systems. A typical MCSoC includes several optimized components integrated together to execute a specific application. Applications range from digital cameras, cellular phones, set-top boxes, PDAs, to biomedical and military instruments.

Different functions in these embedded MCSoCs are typically implemented with software running on a RISC, digital signal processors, or with dedicated hardware IP (Intellectual Property) blocks. These blocks are available from vendors as hard or soft cores. We will discuss later in this chapter how to select these IP cores to build power and performance-efficient multicore systems. The availability of various IP cores with different performance and complexity from many existing providers makes the selection not easy. In addition, selecting suitable cores depends also on the available power, area, and cost budgets. Therefore, designer must be careful and aware about all these factors before even thinking about higher level organization of the target system.

© Springer Nature Singapore Pte Ltd. 2017
A. Ben Abdallah, *Advanced Multicore Systems-On-Chip*,
DOI 10.1007/978-981-10-6092-2_3

Organization of a MCSoC architecture means the software and hardware rela-
tionships between different IP blocks (including on-chip/off-chip memory) and the
interconnection network which links these IP cores together in an efficient manner,
such that several design and performance constraints are satisfied. The hardware
and software design teams should be also aware about the real-time performance
requirement of the system being designed. This is very important because generally
a real-time system has more design constraints than a general multicore system. Consequently, the design of real-time MCSoCs is much more complex that the design of
a general embedded systems.

Modern MCSoC organization guidelines include separation between computation
and communication and between functions and architectures. The applications that
need to run on these MCSoCs have become increasingly complex and have very
tight power and performance requirements. Thus, achieving a satisfactory design
quality under these circumstances is only possible when both communication and
computation refinements are performed efficiently.

As we stated earlier, MCSoCs can be homogeneous or heterogeneous systems.
The organization of each category is of course not similar. The main difference is in
the type and computation power of integrated IP cores. Figure 3.1 shows a general
view of a typical modern MCSoC organization and Fig. 3.2 shows an example of
an embedded multicore system of a typical digital still camera device. The reader
should be also aware that a number of programmable MCSoC platforms are now
commercially available, such as Cell from IBM, Nomadik from STMicroelectronics,
and many others.

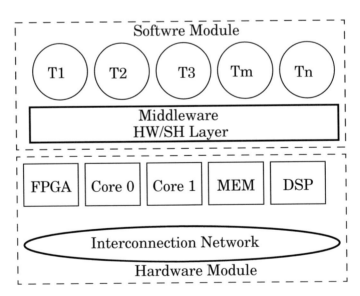

Fig. 3.1 General organization view of a modern typical MCSoC

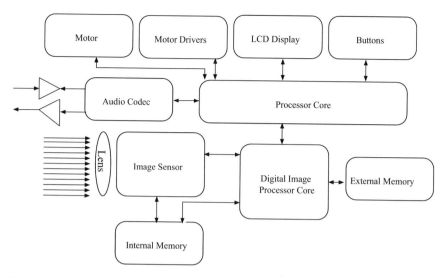

Fig. 3.2 Example of an embedded multicore system for a typical digital still camera

To let the reader first get the "big picture" of such MCSoC system, we will explain in the next part of this section the two main MCSoC categories—homogeneous and heterogeneous. In this chapter, we only focus on the main building blocks of MCSoC system. The system which we assume here is generic and not restricted to a specific kind of embedded applications. The reason is that most building blocks, such as on-chip memory, microprocessor(s), peripheral interfaces, I/O logic control, data converters, and other components are found in most embedded applications. For example, the single chip phone, which has been introduced by several semiconductor vendors, is an example; it includes a modem, radio transceiver, a multimedia engine, security features, and power management functionality all on the same chip.

3.1.1 Heterogeneous MCSoC

A heterogeneous MCSoC is a single chip which combines different cores having different instruction set architectures (ISAs) and computing power interconnected with a sophisticated network or simple shared medium to efficiently link all components together.

Application designers or high-level compilers can choose the most efficient IP cores for the type of processing needed for a given application task.

The main motivation of these systems is that many applications, such as MPEG-2 encoder (see Fig. 3.3), have more than one algorithm during their execution life. This means, a given application has different operations, different memory access patterns, and different communication bandwidth at different execution periods.

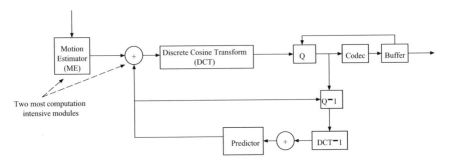

Fig. 3.3 Example of MPEG-2 encoder for a heterogeneous MCSoC system

Another example is in the advanced safety automobile devices, where multiple applications, consisting of several tasks, are executed simultaneously. Each task, invoked by applications, such as image processing, recognition, control or measurement, is assigned to a single processor core. Heterogeneous MCSoCs provide the best performance/power efficiency trade-offs and are a natural choice for embedded systems. The heterogeneous cores increase performance by dividing the work among well-matched cores. This requires many CPU cores for general purpose processing as well as several SIMD processor cores to accelerate specific performance-critical processing. The heterogeneous SoC also can save energy almost at all levels (device, circuit, and logic) of abstraction. In addition, these systems generally use irregular memory and irregular interconnection networks that also save power by reducing the loads in the whole network.

Figure 3.4 shows an example of a heterogeneous MCSoC organization. The above system integrates several typical cores (RISC, accelerators, VLIW, SIMD, etc.) which are found in most modern heterogeneous MCSoC systems. The different cores are

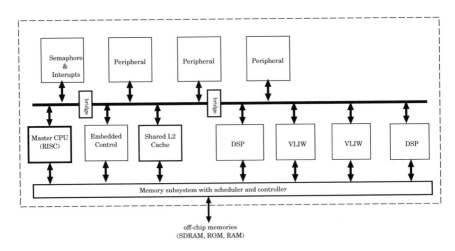

Fig. 3.4 Heterogeneous MCSoC organization example

generally connected to a common pipelined bus (single or multi-layer) with a cache coherence mechanism (discussed later), such as the well-known modified, exclusive, shared or invalid (MESI) protocol [1].

The embedded L2 cache, internal I/O, synchronous dynamic random access memory (SDRAM) are all connected to the bus. The SIMD core is generally a highly specialized parallel processor and is used to process large amount of data, such as images. Additionally, a cache memory is shared by the CPU cores to reduce internal bus traffic and access to the main slow DRAM memory.

3.1.2 Homogeneous MCSoC

An alternative to the previously discussed system is called homogeneous MCSoC. This system is typically built with the same programmable building blocks instantiated several times. This alternative model is often referred in the literature to as parallel architecture model. Parallel architectures were particularly studied in computer science and engineering during the past 40 years. Nowadays, there is a growing interest for such approaches in embedded systems. Figure 3.5 illustrates an example of a typical homogeneous MCSoC organization example. The basic principle of an architecture that exhibits parallel processing capabilities relies on increasing the number of physical resource in order to divide the execution time of each resource.

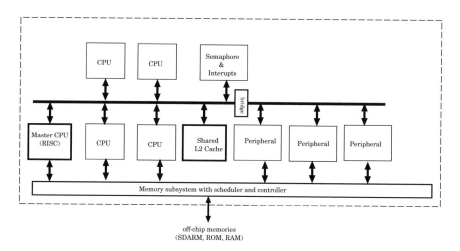

Fig. 3.5 Homogeneous MCSoC organization example

3.1.3 Multicore SoC Applications

As with general architectures, MCSoCs are mainly driven by performance requirements of applications. Therefore, knowing the target application(s) of the system before starting the design is important not only for the selection of appropriate PEs, but also for reducing the overall cost of the system.

There are four well-known applications for MCSoC systems: (1) wireless (2) network, (3) multimedia, and (4) mobile applications. The remaining of this section describes these applications and give some examples for MCSoCs designed for these applications.

Wireless Applications: In this class of applications, MCSoCs are mainly used as wireless base stations (i.e., Luceny Daytona [2]) in which identical signal processing is performed on a number of data channels. Daytona is a homogeneous system with four SPARC V8 CPU cores attached to a high-speed split-transaction. Each CPU has an 8-KB 16-bank cache and each bank can be configured as instruction cache, data cache or scratchpad. The cores share a common address space (see Fig. 3.6).

Network Applications: In this second class, MCSoC can be used as a network processor for packet processing in off-chip networks. The C-5 processor is an example of network processor [3]. In this system, packets are handled by channel cores that are grouped into four clusters of four units each. The traffic of all cores is handled by three buses. In addition to the channel cores, there are also several specialized cores. The executive processor core is a RISC architecture.

Multimedia Applications: Multimedia applications implemented on consumer electronics devices span a vast range of functionality, from audio decoder such as MP3 via video decoder such as H.264 up to advanced picture quality processing such as frame rate up-conversion and motion accurate picture processing (MAPP). Hybrid TV solutions are a very good example because they are virtually capable of executing any of these multimedia applications.

Mobile Applications: The fourth class of MCSoCs application is in the mobile cell phone. Earlier cell phone processors performed base-band operations, including both communication and multimedia operations. As an example, the Texas Instruments' OMAP architecture has several implementations. The OMAP 5912 has two CPU cores: an ARM9 and a TMS320C55x digital signal processor (DSP). The ARM core acts as a master and the DSP core acts as a slave that performs signal processing operations. Another example was implemented by STMicroelectronics and is called Nomadik [4]. It uses an ARM926EJ as its host processor. The ARM926EJ-S processor core runs at up to 350 MHz in 130 nm CMOS process and up to 500 MHz in 90 nm CMOS. The core includes on-board cache, Java acceleration in hardware, and strong real-time debug support Namdik systems are aimed at 2.5 and 3 G mobile phones, personal digital assistants, and other portable wireless products with multimedia capability.

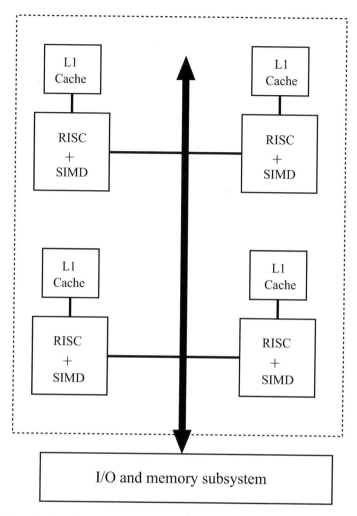

Fig. 3.6 Example of MCSoC application in wireless communication: Lucent Daytona MCSoC

3.1.4 Applications Mapping

As discussed above, today's MCSoC architectures are composed of commercially of-the-shelf available IP blocks. Ultimately, we would like to design a generic heterogeneous MCSoC architecture that is flexible enough to run different applications. However, mapping an application to such heterogeneous SoC is more difficult compared to mapping to a homogeneous one. Today, general practice is to map applications to the architecture at design-time or run-time. Run-time mapping offers a number of advantages over design-time mapping. It mainly offers the following possibilities:

- To avoid defective parts of a SoC. Larger chip area means lower yield. The yield can be improved when the mapper is able to avoid faulty parts of the chip. Also aging can lead to faulty parts that are unforeseeable at design-time.
- To adapt to the available resources. Only at runtime the available resources are known to the mapping algorithm. In addition, the available resources may vary over time for example due to applications running simultaneously or adaptation of algorithms to the environment.
- To enable upgrades of the system.

The objective of the runtime mapping is to determine at runtime a near-optimal mapping of the application to the architecture using the library of process implementations and the current status of the system.

The mapping of the functional subsystems onto SoC hardware resources maybe based on a number of considerations:

- *Support*: support of industry standards. This is very important for processor cores that are programmed by the designers. Generally, industry standard CPU cores have extensive tool chain and library support that eases the application design and debug.
- *Performance*: computationally intensive algorithms such as HD H.264 decoder cannot be implemented effectively on a general purpose processor because of the computational complexity. Instead a function-specific HW core is needed.
- *Flexibility*: evolving standards require flexibility in implementations so that new codecs can be added without the need for a new SoC. This reduces cost and time.
- *Re-usability*: implementation, integration implementation and verification are time consuming tasks and sometimes it is appropriate not to implement a function on the most optimum SoC HW resource in order to make it reusable in future SoCs0.

3.2 MCSoC Building Blocks

As we mentioned in Chap. 1, a typical MCSoC is composed of several components: memories, processing elements, input/output subsystem, and communication subsystem. In most of these MCSoC systems, the cores have separate L1 caches, but share a L2 cache, memory subsystem, interrupt subsystem, and peripherals. Figure 3.7 illustrates a simplified block diagram of a typical MCSoC architecture having different building blocks.

Figure 3.8 shows a general view of a state-of-the-art MCSoC system based on NoC interconnection. In NoC interconnection, PEs communicate with each other using packets and not messages as with shared bus. We will explain this important interconnection paradigm in more detail in Chaps. 4 and 5.

Although systems which are built with NoC approach are scalable and power efficient, the design of such systems is not easy when compared with the design of systems based on shared buses. The reason for such complexity is that the designer must care not only about the computational (PEs) part, but also he must care about the

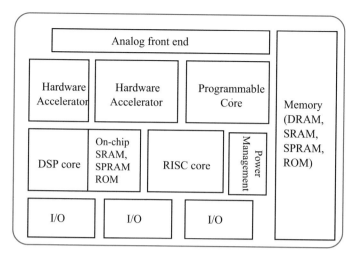

Fig. 3.7 Simplified view of a typical MCSoC architecture with different core and memory types

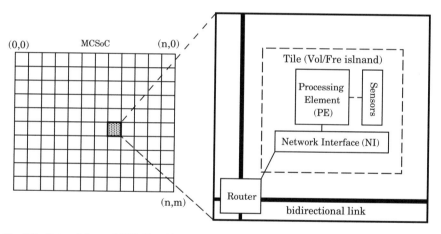

Fig. 3.8 State-of-the-art MCSoC architecture based on network-on-chip paradigm

communication part (how to route packets). In particular, designers must carefully select appropriate topology, routing scheme, control flow, and network interface (NI). Among these blocks, the NI is complex and very important component which must be carefully designed.

The PEs type and computation power depend on the application context and requirements. As we explained in Sect. 3.1, we distinguish two types of architectures: (1) heterogeneous MCSoCs and (2) homogeneous MCSoCs. Heterogeneous systems are composed of different IPs, such as processors, memories, accelerators, and peripherals.

Homogeneous system is a system where the same tile is instantiated several times. Beyond its hardware architecture, MCSoC is generally running a set of software applications divided into tasks and an operating system devoted to manage both hardware and software through a middle-ware layer. Figure 3.1 shows a general view of a MCSoC and the interfacing between the software and hardware modules.

3.2.1 Processor Core

The type and the computation power of the processor core which is embedded in a given MCSoC depend on the target application and whether the core is used for control purpose (master) or computation purpose (slave). Figure 3.9 shows the pipeline stages of a typical RISC processor core. The stages are: fetch, decode, execute, memory access, and write-back stages.

3.2.2 Memory

In a MCSoC, several masters communicate with a single or at most few DRAM (dynamic RAM) memory slaves. The DRAM memory subsystem consists of a memory scheduler, a memory controller, and the DRAM memory. The scheduler arbitrates between multiple requests, whereas controller takes care of bit-level protocol of the DRAM device and activates refreshes, etc. In some design, a sophisticated scheduler reorder the requests such that the DRAM's efficiency is maximized by means of high page hit rate, and low read–write direction turnaround. Figure 3.10 shows an example of a MCSoC based on network-on-chip interconnection network with a single external DRAM memory [5, 6]. In this example, the additional latency added by the router may become a real problem if the memory is highly utilized. That is,

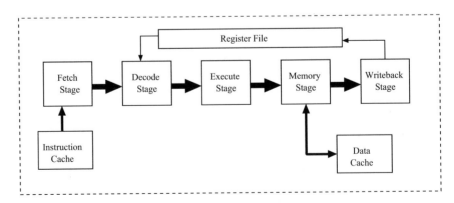

Fig. 3.9 Typical 5 pipeline stages of a RISC processor core

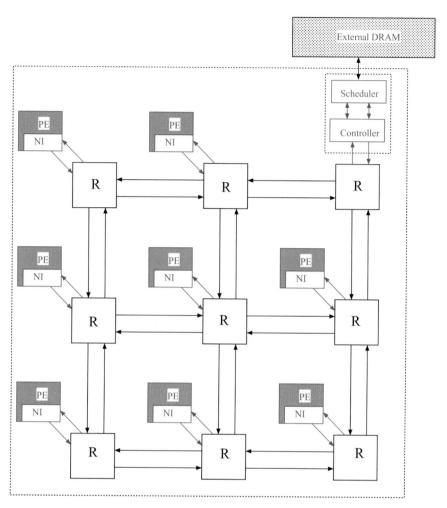

Fig. 3.10 Example of MCSoC with single external DRAM memory

there is a high traffic between the external DRAM and one or more PEs within the system. Such traffic scenario is called traffic hotspot, which affects large portion of the network because blocked traffic reserves many routers and links.

In addition, the requirement to refresh the DRAM at some regular periods reduces the total achievable bandwidth. Moreover, the efficiency is dependent upon the type of transactions, and the address patterns that are presented to the DRAM.

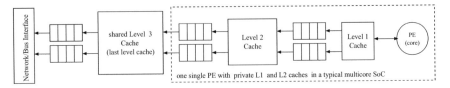

Fig. 3.11 Cache organization in a single node of a typical MCSoC

3.2.3 Cache

Most multicore systems today have one or two levels of dedicated private caches, backed up with a shared last level cache (LLC). The performance and power consumption of a MCSoC is strongly dependent on the performance of the LLC because the LLC can help reduce off-chip memory traffic and contention for memory bandwidth. Figure 3.11 shows three levels of caches in a single node of a typical MCSoC. Cache is efficient because of a program property called *Locality*. The locality says that if a program accesses a particular memory address, it is likely that the next few accesses will be to nearby addresses (spatial locality), and also that the same address is likely to be accessed again within a short time (temporal locality). This is true for instruction fetches, and also for data reads and writes.

System designer can take advantage of the locality of references, to create a hierarchical memory with multiple levels of memory of different speed and size. At the top of this hierarchy, we have a fast, but small memory, which is directly connected to the processor core. The memory sizes increase as we move to lower levels of the hierarchy further away from the processor core. In contrary, the speed drops as we move to lower levels of the hierarchy further away from the processor core.

The minimum amount of data transfered between two adjacent memory levels is called a block or line. Although this could be as small as one word, the spatial locality principle suggests that designer should design caches with larger blocks. If the data requested by a given core is found at a memory level, we say that we have a *hit* at that level. If not, we have a *miss* and the request is sent to the next level down and the block containing the requested data is copied at this level when the data is found. The reason for copying the missed block containing the requested data to the cache is that we want to ensure that next time this (or nearby) data is accessed there will be a hit at this level. The memory system in a MCSoC architecture generally consists of a four-level hierarchy: registers, scratchpad, cache, and main memory.

3.2.4 Communication Protocols

For, a given number of cores, the most appropriate interconnection network depends on a combination of factors, including area/power budget, technology, performance objectives and bandwidth requirements. We have to note here that unlike conventional multiprocessors, performance is not necessary maximized by the highest bandwidth interconnect available.

The traditional form of functional interconnect between different cores in a simple SoC is the on-chip bus which is an array of wires with multiple writers under a mutual-exclusion control scheme. Buses are very simple to design and permits the implementation of efficient hardware mechanisms to enforce cache consistency. In addition, bus-based systems have fair throughput as long as the system is small and there are few masters that initiate data transfers. This is the case with single-core SoC devices, where typically only the core and some advanced peripherals can function as bus masters. Typically, IPs are connected to the bus via standardized protocols, such as advanced extensible interface (AXI), device transaction level (DTL), and open core protocol (OCP).

Using large caches, it is possible to reduce the bus traffic produced by each core, thus allowing systems with greater numbers of cores to be built. Unfortunately, capacitive loading on the bus increases as the number of cores is increased. This effect increases the minimum time required for a bus operation; thus reducing the maximum bus bandwidth.

Multi-bus solutions have provided a temporary solution for small scale systems. However, for large-scale systems, a better solution is still needed. NoC is the promising interconnection paradigm (discussed in Chaps. 4 and 5) for these complex multi- and many core SoCs. Figure 3.12 shows the evolution chart of on-chip communication interconnects for single and MCSoCs.

3.2.4.1 Packet-Switched On-Chip Interconnects

The PEs integrated within modern and future MCSoC are (will be) mostly interconnected by a packet-switched network also called network-on-chip (NoC) [5]. NoC consists of a network of shared communication links and routers, which connect to the various cores through network interfaces (NIs). These NIs convert between the internal NoC protocol on one side and the core's protocol on the other side. For reasons of compatibility and reuse, the latter is typically one of the standardized bus protocols, such as AXI, DTL, and OCP (Fig. 3.13).

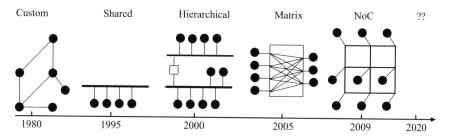

Fig. 3.12 Evolution of on-chip communication interconnect

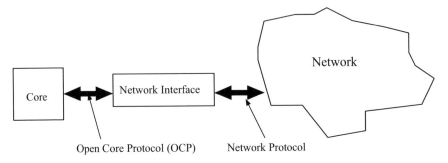

Fig. 3.13 Open core protocol (OCP) and Network protocol (NP) interfacing

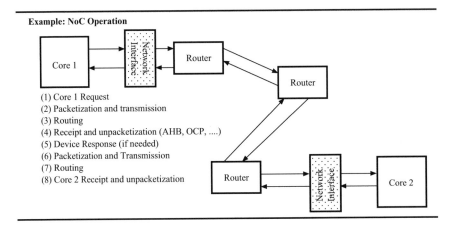

Fig. 3.14 NoC operation

The NI module decouples computation from communication functions. Routers are in charge or routing and are arbitrating the date between the source and destinations PEs through links. Several network topologies have been studied [7, 8]. The NoCs facilitate the design of Globally Asynchronous Locally Synchronous (GALS) property by implementing asynchronous–synchronous interfaces in the NIs. Figure 3.14 shows an example of NoC operation.

3.2.5 Intellectual Property (IP) Cores

An IP core is a block of logic or a software library that we use to design a SoC based on single or multicore. These software and hardware IPs are designed and highly optimized in advance (time to market consideration) by specialized companies and area ready to be integrated with our new design. For example, we may buy a software library to perform some complex graphic operations and integrate that library with our existing code. We may also obtain the above code freely from an open-source site online. Universal asynchronous receiver/transmitter (UARTs), central processing units (CPUs), ethernet controllers, and PCI interfaces are all examples of hardware IP cores.

As essential elements of design reuse, IP cores are part of the growing electronic design automation (EDA) industry trend toward repeated use of previously designed components. Ideally, an IP core should be entirely portable. This means the core must be able to easily be integrated (plug-and-play style) into any vendor technology or design methodology. Of course there are some IPs that are not standard and may need some kind of interface (called wrapper) before integrating it into our design. IP cores fall into one of two main categories: soft cores and hard cores:

(1) Soft IP Core: Soft IP cores refer to circuits which are available at a higher level of abstraction, such as register-transfer level (RTL). These type of cores can be customized by the user for specific applications.
(2) Hard IP Core: A hard IP core is one where the circuit is available at a lower level of abstraction such as the layout-level. For this type of core, it is impossible to customize it to suit the requirements of the embedded system. As a result, there are limited opportunities in optimizing the cost functions by modifying the hard IP.

A good IP core should be configurable so that it can meet the needs of many different designs. It also should have a standard interface so that it can be integrated easily. Finally, a good IP core should come in forms of complete set of deliverables: synthesizable RTL, complete test benches, synthesis scripts, and documentation. The example shown in Fig. 3.15 is for a hardware IP core from Altera FPGA provider [9].

3.2.6 IP Cores with Multiple Clock Domains

The IP cores integrated within a given MCSoC may work at different clock rates. For example, some SoC may have more than three clock domains. In addition, many

---

Example: Altera Intellectual Property

The Altera IP core site (http://www.altera.com/products/ip/ip-index.jsp)
Altera IP site provides access to a wide variety of IP blocks of different size and complexity:

- basic arithmetic blocks to transceivers,
- memory controllers,
- microprocessors,
- signal processing, and
- protocol interfaces

Fig. 3.15 Intellectual property example

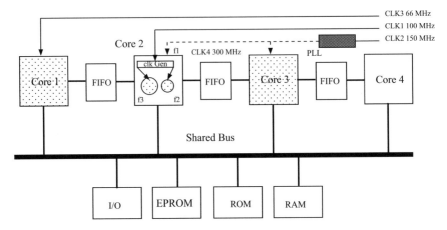

Fig. 3.16 Three clock domains MCSoC

embedded cores operate internally using multiple frequencies. Figure 3.16 shows a
simple design that comprises three cores with three different physical clocks. In this
example, Core 2 consists of three modules operating at different frequencies (f1, f2,
and f3). A physical clock is a chip-level clock; for example, it can come from an
oscillator, or a phase-locked loop (PILL). All the internal clocks generated from the
same physical clock are considered to be a part of the same physical clock domain.

In a MCSoC system, the multi-frequency blocks communicate one with each
other through synchronization logic and/or FIFO memory blocks. Such design has
the advantage of low power and low silicon area. However, the main design difficulties with multi-frequency embedded cores are the clock skew, and synchronization
problems.

3.2.7 Selection of IP Cores

This section investigates the classification and selection of processor cores and considers the MCSoC infrastructure that allows an efficient mix of different types of cores
and function-specific hardware cores to access shared resources in the SoC. Selection
of appropriate IP cores depends on the application mapping output. Depending of
the target application, mapping of the functional subsystems to a MCSoC hardware
resources generally involves the following cores:

- *Host CPU*: The host CPU is generally an industry standard core such as, MIPS and
  ARM CPUs. Typically, these cores have a large application code and thus need to
  access code and data in an external SDRAM memory.
- *VLIW processor*: Generally, this core provides scalability and processing power.
  This processor core exploits fine-grained data parallelism. Code and data segments
  are typically large, so the VLIW processor core also needs access to SDRAM.

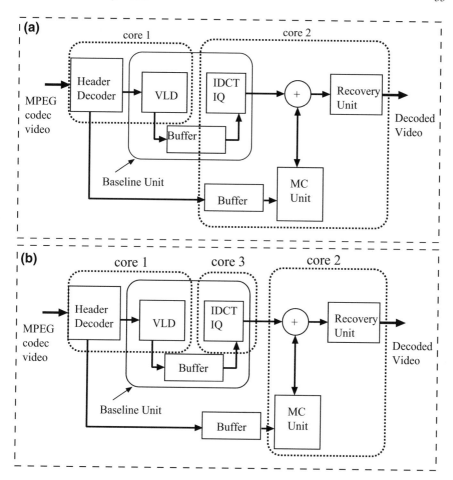

Fig. 3.17 Example of mapping of an MPEG-2 decoder. **a** Using two cores, **b** Using three cores

- *Embedded control CPU*: Small to medium sized code base. Architectural and commercial reasons often use processor core from the same processor provider as the host CPU. For architectural consistency, such core is connected to the same bus structures as the host CPU or VLIW processor cores.
- *Fixed point DSP*: generally deeply embedded into the MCSoC architecture. Often, DSP cores are connected into the SoC infrastructure via HW semaphore mechanisms.
- *Function-specific HW core*: massively parallel computation core that makes use of fine-grained parallelism as well as coarser parallelism and typically processes large data sets. Some cores are connected to external real-time interfaces, and therefore need real-time performance.

As an example of mapping, a given application to different cores within a MCSoC system, consider an example of an MPEG-2 decoder application which consists of a baseline unit, a motion compensation (MC) unit, a recovery unit, and the associated buffers. The baseline unit consists of a VLD (variable length decoder), an IQ/ IZZ (inverse quantization/inverse zigzag) module, IDCT (inverse discrete cosine transform) modules, and the buffer. Figure 3.17 shows this application running on a multicore system with two or three cores.

3.3 MCSoC Memory Hierarchy

The memory architecture of an embedded MCSoCs strongly influences area, power, and performance of the entire system. In these systems, more on-chip silicon is devoted to memory than to anything else on the chip. This requires special attention that must be dedicated to the on-chip memory organization.

The memory organization of embedded MCSoC systems varies widely from one to another, depending on the application and market segment for which the SoC is targeted. Broadly speaking, program memories for MCSoCs are classified into (1) primary memory and (2) secondary memory.

The primary memory is the memory that is addressed by core(s) and holds current data set that is being processed as well as the program (text) code. This memory may consist of main memory typically implemented in DRAM technology, and a hierarchy of smaller and faster caches (SRAMs) or Scratchpad memories (SPRAMs), that hold the copies of some of the data from the main memory.

The secondary memory maybe also used for long-term storage. Embedded MCSoC systems often include flash memory as the secondary storage, e.g., for storing pictures in a digital camera.

In the remaining part of this section, we will discuss in a fair amount of details the many alternatives for on-chip and off-chip memory usage that SoC designers must understand.

3.3.1 Types on On-Chip Memory

There are three broad categories of on-chip memories that system designer can use. The first type is called static random access memory (SRAM). This is quite known memory architecture and is found in almost all type of computers and not only embedded multicore systems. SRAM is very common in SoC designs because it is fast and is built from the same transistors used to build all of the logic on the SoC, so no process changes are required. Further, due to its good characteristics (mainly speed), SRAMs are generally used for caches to solve the processor–memory speed mismatches.

Most SRAM bit-cells require at least four transistors and some require as many as ten; so on-chip dynamic RAM or DRAM is becoming increasingly popular. DRAM stores bits as capacitive charge, so each DRAM bit cell requires only one transistor and a capacitor. DRAM's main advantage is density. But, DARM is slower than SRAM and has some particular requirements that affect system design such as the need for periodic refresh. Further, the capacitors in the DRAM bit-cells require specialized processing, which increases die cost. This is of course not a good situation for strictly cost constrained embedded MCSoC systems.

Every SoC needs memory that remembers code and data even when the power is off. Thus, the cheapest and the least flexible memory is ROM and so is our second type of memory in this discussion. ROMs are not flexible since their contents cannot be changed after the system is fabricated. Fortunately, EPROM and flash memory are good alternatives. Figure 3.18 illustrates a simplified view of a MCSoC architecture with different core and memory types.

A given memory bank can be organized as a single-access RAM or a dual-access RAM to provide single or dual access to the memory bank in a single cycle. Also the on-chip memory banks can be of different sizes.

The good thing for smaller banks is that they consume less power per access than the larger memories. Embedded multicore systems may also be interfaced to off-chip memory, which can include SRAM and DRAM. If the system is targeted for low to medium complex embedded applications, purely SPRAM based on-chip organization is recommended. FIFO memories can be also used to inter-core communication inside the MCSoC chip as shown in Fig. 3.19.

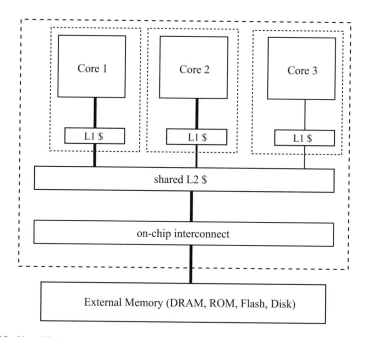

Fig. 3.18 Simplified view of a MCSoC architecture having different memories

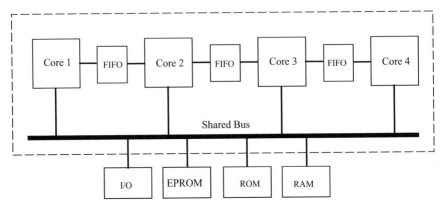

Fig. 3.19 Example of four cores communicating via FIFOs

3.3.2   Scratchpad Memory

Scratchpad memory (SPRAM) is a high-speed internal memory directly connected to the CPU core and used for temporary storage to hold very small items of data for rapid retrieval. Scratchpads are employed for simplification of caching logic, and to guarantee a unit can work without main memory contention in a system employing multiple cores, especially in embedded MCSoC systems. They are suited for storing temporary results.

While a cache memory uses a complex hardware controller to decide which data to keep in cache memories (L1 or L2) and which data to prefetch, the SPRAM approach does not require any hardware support in addition to the memory itself, but requires software to take control of all data transfers to and from Scratchpad memories. That is, it is the responsibility of the programmer to identify data section that should be placed in SPRAM or place code in the program to appropriately move data from on-chip memory to SPRAM. For this reason, SPRAMs are sometimes called "software controlled caches." Figure 3.20 illustrates the memory subsystem architecture with 2 SPARMs (level 1 and level 2).

3.3.3   Off-Chip Memory

When embedded system designers need a large amount of RAM storage, then off-chip DDR (double-data-rate) SDRAM is likely to be the good choice. Even if an embedded design only requires a small fraction of the capacity of a DDR memory chip or module, it may still be more economical to pay for the excess capacity because the system price will still be lower.

Adding a DDR memory port to a MCSoC design creates the need for an on-chip DDR memory controller. In the same way, system design considerations may make

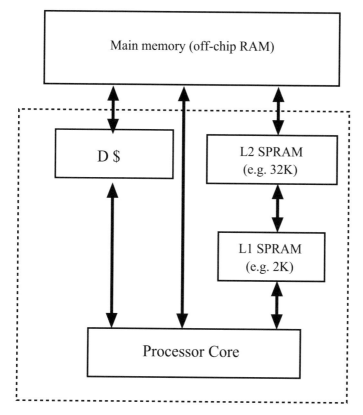

Fig. 3.20 MCSoC memory subsystem with SPARM (only interconnection for one node is shown for simplicity)

it more desirable to have nonvolatile memory reside off-chip. Again, this option is adopted when a large nonvolatile memory is needed or when the manufacturing costs needed to add EEPROM to the MCSoC are expensive due to limited available budget. In this case, the hardware design team should add a Flash memory controller which of course will add some extra hardware and cost to the system.

3.3.4 Memory Power Reduction in SoC Designs

Due to recent increases in VLSI density, SoC designers have exploited the additional silicon available on chips to integrate embedded memories such as SPRAMs, FIFOs, and caches to store data for a large number of cores.

Since these embedded memories are implemented inside the chip, the communication latency is low or even negligible. Thus, they allow for significantly better system performance and lower power compared to a solution where off-chip memories are used.

It was found by several researchers that the memory subsystem accounts for up to 50–70% of the total power consumption of the system [10]. This reflects the importance of limiting the energy consumption of memory subsystem. One possible architectural approach for memory energy reduction is the replacement of traditional cache-based memory subsystem by customized SPRAM based one.

The energy savings from this solution comes from the fact that SRPARM consumes less energy per access than a cache due to the absence of additional hardware (e.g., tag memory) present in a cache.

With more transistors becoming available on chip, the percentage of area taken by memory is increasing. In addition to the power projection, the ITRS 2003 report projects that in 2012 memory will occupy about 90% of a chip. This means that only about 10% will be left for the processor's computing blocks. Figure 3.21 shows the projection of memory/logic composition of a power-constrained SoC chips.

As we mentioned earlier, most memories embedded in MCSoCs use SRAM technology. The key sources of power consumption in such memories are:

- Static or leakage power dissipated by the logic in the periphery and memory array.
- Dynamic or switching power dissipated when read or write operations are performed.

The dynamic power consumed by a memory when a read or write operation occurs, can be divided into the power consumed by the following components:

- Toggling of the clock network
- Registers for data/address latching on memory I/Os
- Bit-lines in the memory array
- Peripheral logic to decode the address
- Core memory cells changing state.

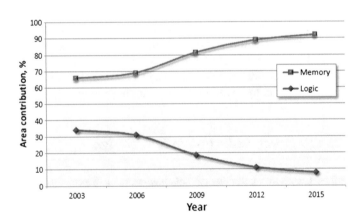

Fig. 3.21 Projection of memory/logic composition of power-constrained SoC chips [11]

3.4 Memory Consistency in Multicore Systems

In the traditional Von Neumann machines, instructions appear to execute in the order specified by the programmer or compiler regardless if the implementation of the machine actually executes them in a different order. For example, a load instruction should return the last value written to the memory location. Likewise, a store instruction to a memory location determines the value of the next load. All sequential programs assume this strict rule when they are executed on a uni-processor.

Multithreaded programs running on multicore systems complicate both the programming model and the implementation to enforce a given model. More precisely, the value returned by a given load is not clear because the most recent store instruction may have occurred on a different core. Thus, system designers generally define memory consistency models to specify how a processor core can observe memory accesses from other cores in the same system. Serial consistency is a model defined such that the result of any execution is the same as if the operations of all processor cores were executed in some serial order, and the operations of each individual core behave in this sequence in the order specified by its program. The addition of cache memories to these systems affects how such consistency is implemented.

A cache memory allows processor speed to increase at a greater rate than the main memory speed by exploiting what is known as "time" and "space" localities.

The process of connecting memory locations with cache lines is called "mapping." Since cache is smaller than the main memory, the same cache lines are shared for different memory locations. Each cache line has a record of the memory address called tag. This tag is used to track which area of memory is stored in a particular cache line.

The way these tags are mapped to cache lines can have a beneficial effect on the way a program runs. Caches can be organized in one of several ways: direct mapped, fully associative, and set associative. Figure 3.22 shows an example of direct-mapped cache organization. Cache operations are mainly done in hardware and their operation is all hardware-based and automatic from a programmer's point-of-view. In other words, details of the cache hierarchy do not affect the instruction set architecture of the processor. While caches do not present a real problem in a uni-processor system, they considerably complicate memory consistency for systems designed with multi- and many cores. This problem is known in the literature as *cache coherence* problem.

3.4.1 Cache Coherence Problem

In a single-core system, the coherence problem appears when an I/O peripheral bypasses the cache on the system bus and flows directly to and from the main memory (DRAM). This problem can be easily solved by software (compiler) because the single-thread context imposes a well-defined thread order and the software is always informed on each trap and interrupt caused by a given I/O. The compiler, then, tags

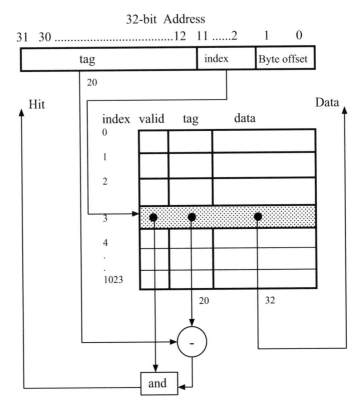

Fig. 3.22 Direct-mapped cache organization

data as cacheable and non-cacheable. Only read-only data is considered cacheable and put in private cache. All other data are non-cacheable, and can be put in a global cache, if available.

In multicore-based systems things are quite different and more serious because it is difficult to keep record about the order of instructions in different threads running in simultaneously and in different processor cores.

This "coherence" problem comes from the multiple copies of the same memory location, not only in the cache hierarchy, but also in more low-level hardware buffers for memory accesses inside the processor core. The coherence problem here is more difficult to solve than in single core system because the software is not always informed and on-chip communication patterns are not clearly seen by the system's software.

Figure 3.23, illustrates an example of cache coherence problem. As shown in the above figure, the value returned by a given load is not clear because the most recent store may have occurred on a different core. We have to note here that this problem is not very different from multiprocessor (multiple chips) cache coherence problem. Thus, system designers generally define memory consistency models to specify how

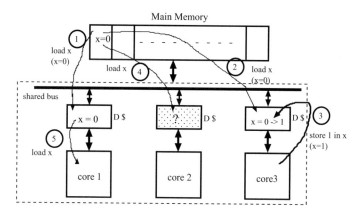

Fig. 3.23 Cache coherence problem example without coherence protocol

a processor core can observe memory accesses from other processor cores in the system.

A multicore system is said to be *cache coherent* if the execution of a given program leads in a valid ordering of reads and writes to a memory location.

3.4.2 Cache Coherence Protocols

We have to note first that the solution for cache coherence problem is a general problem with multiprocessors and only limited to multicore systems or MCSoCs. There exist many coherence algorithms and protocols.

For a small-scale bus-based system, snooping bus is generally used. There are two basic methods to utilize the inter-core bus to notify other cores when a core changes something in its cache. One method is referred to as *update*. In the update method, if core 1 modifies variable 'y' it sends the updated value of 'y' onto the inter-core bus. Each cache is always listening (snooping) to the inter-core bus so that if a cache sees a variable on the bus which it has a copy of, it will read the updated value. This ensures that all caches have the most up-to-date value of the variable.

Another method which utilizes the inter-core bus is called *invalidation*. This method sends an invalidation message onto the inter-core bus when a variable is changed. The other caches will read this invalidation signal and if its core tries to access that variable, it will result in a cache miss and the variable will be read from main memory.

The update method causes a significant amount of traffic on the inter-core bus because the update signal has to be sent onto the bus every time the variable is updated. However, the invalidation method only requires that an invalidation signal is sent for the first time a variable is altered; this is why the invalidation method is the preferred method. Table 3.1 shows all cache coherence states.

Table 3.1 Cache coherence states

State	Permission	Definition
Modified (M)	Read, write	All other caches in I or NP
Exclusive (E)	Read, write	The addressed line is in this cache only.
Owned (O)	Read	All other caches in S, I, or NP
Shared (S)	Read	All other caches in M or E
Invalid (I)	None	None
Not Present (NP)	None	None

MSI (Modified, Shared, and Invalid) Protocol: MSI is a basic but well-known cache coherency protocol. These are the three states that a line of cache can be in. The *Modified* state means that a variable in the cache has been modified, and therefore has a different value than that found in main memory. The cache is responsible for writing the variable back to main memory. The *Shared* state means that the variable exists in at least one cache and is not modified. The cache can evict the variable without writing it back to the main memory. The *Invalid* state means that the value of the variable has been modified by another cache and this value is invalid.

MESI (Modified, Exclusive, Shared, and Invalid) Protocol: Another well-known cache coherency protocol is the MESI protocol. The modified and invalid states are the same for this protocol as they are for the MSI protocol. This protocol introduces a new state; the exclusive state. The exclusive state means that the variable is in only this cache and the value of it matches the value within the main memory. This now means that the Shared state indicates that the variable is contained in more than one cache.

MOSI (Modified, Owned, Shared, and Invalid) Protocol: The MOSI protocol is identical to the MSI protocol except that it adds an Owned state. The Owned state means that the processor "Owns" the variable and will provide the current value to other caches when requested (or at least it will decide if it will provide it when asked).

3.4.2.1 Directory-Based Cache Coherency

The snooping protocol works well with system based on shared bus (natural broadcast medium). However, large-scale MCSoC (and multiprocessors) may connect cores/processors with memories using switches or some other kind of complex interconnects. Thus, a new method is needed.

The alternative for the "snoopy-bus" scheme is a protocol known as "directory" protocol [12, 13]. The basic idea in this scheme is to keep track of what is being shared in one centralized place called directory. This method scales better than snoopy-bus. In this approach, each cache can communicate the state of its variables with a single directory instead of broadcasting the state to all cores.

Cache coherence protocols guarantee that eventually all copies are updated. Depending on how and when these updates are performed, a read operation may sometimes return unexpected values. Consistency deals with what values can be returned to the user by a read operation (may return unexpected values if the update is not complete). Consistency model is a contract that defines what a programmer can expect from the system.

3.5 Chapter Summary

Increasing processing power demand for new embedded consumer applications made the convectional single-core-based designs no longer suitable to satisfy high performance and low power consumption demands. In addition, continuous advancements in semiconductor technology enable us to design a complex multicore systems-on-chip (MCSoCs) composed of tens or even hundreds of IP cores.

Integrating multiple cores on a single chip has enabled embedded system hardware designers to provide more features and higher processing speeds using less power, thus solving many design problems. However, no thing is really free! The designer of these embedded MCSoCs is no longer dealing with the familiar homogeneous and symmetric multiprocessing (SMP) model of large computer systems. Rather, he may have dozens or hundreds of processor core to program and debug, a heterogeneous and unbalanced mix of DSP, RISC, IPs and complex on-chip network architectures, operating asymmetrically. This is not an easy task.

In this chapter, we tried to explain the main components of a typical MCSoC system. The goal is to give a clear idea about the architecture and function of the main building blocks. In the next two chapters, we will describe in details the network-on-chip interconnection which is a promising on-chip interconnection paradigm for future multi- and many core SoCs.

References

1. M.S. Papamarcos, J.H. Patel, A low-overhead coherence solution for multiprocessors with private cache memories, in *ISCA '84 Proceedings of the 11th Annual International Symposium on Computer Architecture* (1984), pp. 348–354
2. J. Knobloch, E. Micca, M. Moturi, C.J. Nicol, J.H. O'Neill, J. Othmer, E. Sackinger, K.J. Singh, J. Sweet, C.J. Terman, J. Williams, A single-chip, 1.6-billion, 16-b MAC/s multiprocessor DSP. IEEE J. Solid-State Circuits **35**(3), 412–424 (2000)
3. C-5 Network Processor Architecture Guide, C-Port Corp., North Andover, MA, 31 May 2001
4. STMicroelectronics: http://www.st.com/internet/com/home/home.jsp
5. A. Ben Abdallah, M. Sowa, Basic network-on-chip interconnection for future gigascale MCSoCs applications: communication and computation orthogonalization, in *Proceedings of the Joint Symposium on Science, Society and Technology (JASSST2006)* (2006), pp. 1–7, 4–9 December 2006

6. A. Ben Abdallah, T. Yoshinaga, M. Sowa, Scalable core-based methodology and synthesizable core for systematic design environment in multicore SoC (MCSoC), in *Proceedings IEEE 35th International Conference on Parallel Processing Workshops* (2006), pp. 345–352, 14–18 August 2006
7. A.B. Ahmed, A. Ben Abdallah, Graceful deadlock-free fault-tolerant routing algorithm for 3D network-on-chip architectures. J. Parallel Distrib. Comput. **74**(4), 2229–2240 (2014)
8. A.B. Ahmed, A. Ben Abdallah, Adaptive fault-tolerant architecture and routing algorithm for reliable many-core 3D-NoC systems. J. Parallel Distrib. Comput. **9394**, 30–43 (2016)
9. Altera: http://www.altera.com/
10. International Technology Roadmap for Semiconductors, 2005 Edition
11. International Technology Roadmap for Semiconductors, 2003 Edition, System Drivers
12. L.M. Censier, P. Feautrier, A new solution to coherence problems in multicache systems. IEEE Trans. Comput. **c–20**(12), 1112–1118 (1978)
13. D. Chaiken, C. Fields, K. Kurihara, A. Agarwal, Directory-based cache coherence in large-scale multiprocessors. Computer **23**(6), 49–58 (1990)

Chapter 4
Multicore SoC On-Chip Interconnection Networks

Abstract Global interconnects are becoming the principal performance bottleneck for high-performance multicore SoCs. Since one of the main purposes of SoC design is to shrink the size of the chip as smaller as possible while seeking at the same time for more scalability, higher bandwidth and lower latency. Conventional bus-based systems are no longer reliable architecture for SoC due to a lack of scalability and parallelism integration. During this last decade, network-on-chip (NoC) has been proposed as a promising solution for future systems on chip design. It offers more scalability than the shared bus-based interconnection, and allows more processors/-cores to operate concurrently. This chapter presents architecture and design of a two-dimensional NoC system suitable for medium scale multicore SoCs.

4.1 Introduction

Future high-performance embedded SoCs will be based on multi- and manycore approaches with nanoscale technology consisting of hundreds of processing and storage elements. These new paradigms are emerging as a key design solution for today's nanoelectronics design problems. The interconnection structure supporting such systems will be closer to a sophisticated network than to current bus-based solutions. Such network must provide high throughput and low latency while keeping area and power consumption low.

Network-On-Chips (NoCs) [1–4] provide a good way of realizing interconnections on silicon and largely alleviate the limitations of bus-based solutions. Deep submicron processing technologies have enabled the implementation of new application-specific architectures that integrate multiple software programmable cores and dedicated hardware components together onto a single chip. Recently, this kind of architecture has emerged as key design solutions for today's design challenges, which are being driven by various emerging applications, such as wireless communication, broadband/distributed networking, distributed computing, and multimedia computing.

© Springer Nature Singapore Pte Ltd. 2017
A. Ben Abdallah, *Advanced Multicore Systems-On-Chip*,
DOI 10.1007/978-981-10-6092-2_4

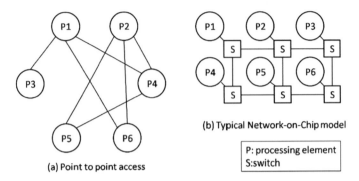

Fig. 4.1 Typical paradigms: **a** circuit switching, **b** packet switching

NoC is a scalable interconnect with a huge potential to handle the increasing complexity of current and future multicore SoCs. In such paradigm, cores are connected
via a packet-switching communication network on a single chip. This scheme is similar to the way that computers are connected to the Internet. The packet-switching network routes information between network clients (e.g., PEs, memories, and custom
logic devices). Figure 4.1 illustrates a point-to-point network and a Network-on-Chip
model.

Packet switching approach supports asynchronous data transfer and provides
extremely high bandwidth by distributing the propagation delay across multiple
switches and effectively pipelining the packet transmission. In addition, it offers
several other promising features. First, it transmits packets instead of words. As a
result, dedicated address lines, like those used in bus-based systems, are not necessary since the destination address of a given packet is included in the packet's
header. Second, transmission can be conducted in parallel if the network provides
more than one transmission channel between a sender and a receiver. Thus, unlike
bus-based systems, NoC presents theoretical infinite scalability, facilitates IP cores
reusing, and has higher level of parallelism. A NoC architecture, named OASIS NoC
(ONoC), was developed in [5–9]. The above network is based on mesh topology [10]
and uses wormhole like switching, a first-come-first-served (FCFS) scheduler, and
retransmission flow control similar to conventional ACK/NACK flow control.

NoC research issues include a lot of trade-offs such as topology, routing, switching, scheduling, flow control, buffer size, packet size, and any optimization techniques. It is difficult to analyze these parameters using only high-level simulations.
Therefore, NoC prototype is an essential design phase for evaluating the performance
of the NoC architectures under real applications [11].

The remaining of this chapter presents architecture and design details of a 2D-
mesh NoC architecture. The chapter also describes a so called short pass link (SPL)
to optimize mesh topology by reducing (in some cases) the number of hopes between
nodes experiencing heavy traffic.

4.2 Network-on-Chip Architecture

As we earlier stated, an NoC architecture is generally characterized by its topology, routing, switching, flow control, and arbiter techniques. There are various trade-offs when selecting these parameters. So, designers need to take care and deeply understand about all the design choices.

4.2.1 Topology

The topology defines the way routers and links are interconnected. Topology is an important design choice as it defines the communication distance and its uniformity. Some of the most used topologies are depicted in Fig. 4.2.

The choice of a topology depends on its advantages and drawbacks. Usually, regular topologies (Fig. 4.2a–e) are preferred over irregular ones (Fig. 4.2f), because of their scalability and reusable pattern. Otherwise, irregular or mixed topologies can be more conveniently be adapted to specific needs of the application. This depends on the target application which may require some area, power, or timing constraints that need to be strictly satisfied. In this case, regular topologies might not be the right approach to implement such special applications, and custom irregular ones offer better flexibility to meet the desired requirements. On the other hand, one of

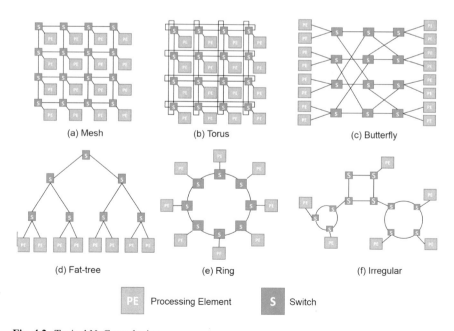

Fig. 4.2 Typical NoC topologies

the main problems that irregular topologies suffer from is the design time needed to profile the application and decide the best topology layout that satisfies these design requirements.

The mesh and torus-based topologies are considered as the most commonly used on-chip network topologies. Together they constitute over 60% of 2D-NOC topology cases [12]. Mesh and Torus are depicted in Fig. 4.2a, b respectively. Both of them can have four neighboring connections; but, only Torus has wraparound links connecting the nodes on network edges. Other topologies like butterfly, fat-tree, and ring (depicted in Fig. 4.2c–e, respectively) have roughly even proportion.

Compared with other on-chip network topologies, the mesh topology in particular can achieve better application scalability. The implementation of routing functions in mesh topology is also simpler and can be characterized well. In the on-chip interconnection networks for on-chip multiprocessor systems, the mesh architecture is widely used and preferable. An example of on-chip multicore system that uses mesh topology is Intel-Teraflops system. The 80 homogeneous computing elements are interconnected through NoC routers in the 2D mesh 8×10 network topology.

The routing methods are selected depending on the topology [13, 14]. Routing algorithm can be easily implemented in a standard topology (e.g., Mesh, Torus, Star) because each router sends the same routing path to its neighboring nodes. However, for customized topology, routing is generally more difficult and it is necessary to design specific routing mechanism. Thus, the design time may be longer than standard [15, 16]. Figure 4.3 illustrates an example of a 3×3 mesh-based NoC system.

Fig. 4.3 Example of a 3×3 NoC based on mesh topology. R: router/switch, PE: processing element, NI: network interface

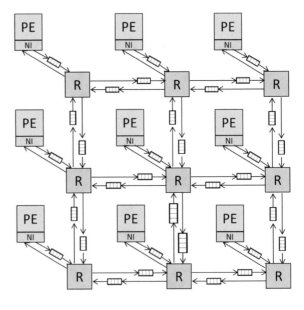

4.2.2 Switching

There exist two main types of switching methods in NoC interconnects: (1) circuit switching and (2) packet (or flit) switching. In the first method, the path between a given source and destination pair should be first established and reserved before starting to send the actual data. This offers some performance guarantees as the message is sure to be transferred to its destination without the need for buffering, repeating, or regenerating. Moreover, if during the establishment of the path a problem is detected (such as failure or high congestion), the source node can recompute another safer path to be reserved again. However, the path setup required for each message increases the latency overhead, in addition to the extra congestion caused by the different control data traveling the network and competing with the actual data for the network resources. Therefore, it is best suited for predictable transfers that are long enough to amortize the setup latency.

Packet-switching is more common and it is widely used in NoC systems. In packet switching, routers communicate through transmitting packets/flits through the network. The transmission of a given packet should not block the communication of other ones in the network. To solve this problem, a forwarding method (switching policy) can be selected to define how the network resources (link and switched) are reserved and how they are torn down after the transfer completion. The forwarding methods have a big impact on the NoC performance and each one of them has its advantages and drawbacks. In packet switching, store-and-forward (SF), wormhole (WH), and virtual-cut-through (VCT) are considered as the main switching methods.

4.2.2.1 Store-and-Forward (SF)

In this switching method, each message should be divided into several packets. As depicted in Fig. 4.4, each packet is completely stored in a FIFO buffer before it is forwarded to the next router. Therefore, the size (depth) of FIFO buffers in the router is set similar to the size of the packet in order to be able to completely store the packet. This represents the main drawback of this switching policy since it requires a significant amount of buffer resources which increases as we increase the packet size. This amount of allocated buffer slots has a huge impact on the area and power consumption of the NoC system. Moreover, as can be seen in Fig. 4.4, node (0,2) has two empty slots since the first two flits of Packet-4 (P4F1 and P4F2) have been already transmitted. Despite the available two slots, Packet-5 (P5) in node (0,1) is still stalled. This is because in order to be forwarded, all the four slots in node (0,2) should be freed; therefore, P5 can be forwarded only when P4 is forwarded as well and the buffer slots are freed. Store-and-Forward was the first switching method that has been used in many parallel machines [17–19]. It was also in the first prototypes and designs of NoC [20–24].

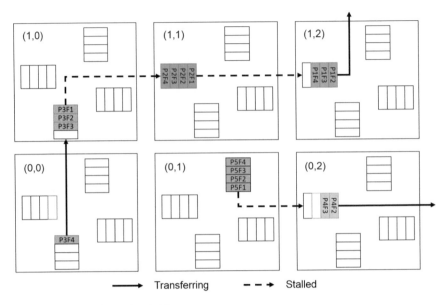

Fig. 4.4 Store-and-forward switching

4.2.2.2 Wormhole (WH)

Wormhole switching (WH) is one of the most popular, well-used and well suited for NoC systems. In WH switching method, represented in Fig. 4.5, packets are divided into a number of flits. As can be seen in Fig. 4.5, the four flits of Packet-1 (P1F1, P1F2, P1F3, and P1F4) are dispersed in four different routers. Therefore, no need for buffer resources to host the entire packet. The main advantage of the wormhole switching is that the buffer size can be set as small as possible to reduce the buffering area cost. This responds to the area and power overhead of SF. However, blocking is one of its major drawbacks. As depicted in Fig. 4.5, the last flit of P1 is located in the head of the south input-buffer of node (1,0).

In the tail of the same input-buffer, the first flit of Packet-2 (P2) is requesting the grant to be forwarded to the north output-port (heading for node (2,0)). In this scenario there is a tight dependency between the first P1F4 and the second P2F1. In other words, if P1F4 is forwarded then P2F1 can be forwarded as well; however, in case where P1F4 is blocked for congestion or failure reasons in the downstream nodes, then P2F1 is blocked too. Consequently, the remaining flits of P2 and the dependent other flits will be blocked as well. This will lead to the partial or entire system deadlock and a significant performance degradation. One of the solutions, to solve this problem in WH switching, Virtual-channels [25] can be used. This will be discussed later in this chapter (Sect. 2.1.5).

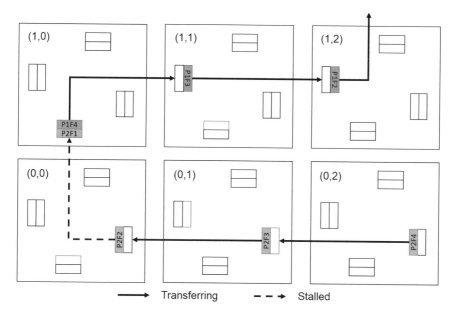

Fig. 4.5 Wormhole switching

The wormhole switching method was first introduced in [26]. The work in [27] has presented also the performance of the wormhole switching in k-ary n-cube interconnection networks.

4.2.2.3 Virtual-Cut-Through (VCT)

Figure 4.6 demonstrates Virtual-Cut-Through (VCT) switching. VCT is an intermediate forwarding method that has the properties of both SF and WH. As represented in Fig. 4.6, with VCT it is possible to forward flits one after another. So, flits from different packets can share the same input-buffer eliminating the stalling caused by SF. In order to solve the blocking problem found in WH switching, VCT requires that the buffer depth should be equal to the packet size (number of flits in the packet). This buffer size is needed to store blocked flits.

When blocking happens, flits are stored in a router next to the blocked one. The buffer size is larger than WH switching since the entire packet is stored. However, the forwarding latency is much smaller than SF switching. This is because in the Store-and-Forward packet switching method the packet is completely stored before it is forwarded to the next router and the delay to wait for the complete packet storing is very long.

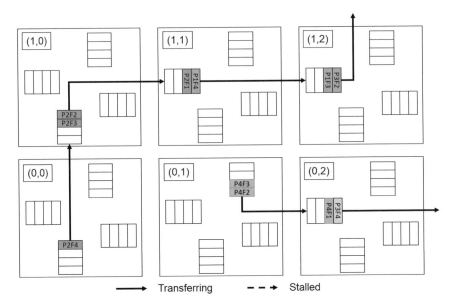

Fig. 4.6 Virtual-cut-through switching

4.2.3 Flow Control

Flow control determines how resources, such as buffers and channels bandwidth
are allocated, and how packet collisions are resolved [1]. Whenever the packet is
buffered, blocked, dropped, or misrouted, this depends on the flow control strategy.
A good flow control strategy should avoid channel congestion while reducing the
latency. ON/OFF, credit-based, and ACK/NACK are commonly used control flows
used in NoC and are explained in this subsection.

4.2.3.1 ON/OFF Flow Control

ON/OFF flow control [28] has protocols which can manage data flow from upstream
routers while issuing a minimal amount of control signals. It is able to do this because
it has only two states: ON or OFF. This control flow has threshold values, which are
dependent on the number of free buffers in downstream routers. The threshold values
are used to decide the states of the control signals. When the number of free buffers
is over the threshold, downstream routers emit an OFF signal to upstream routers,
stopping the flow of flits. Meanwhile, the downstream routers send flits to other nodes,
and the number of free buffers becomes less than the threshold value, downstream
routers emit an ON signal to upstream routers, restarting the flow of flits. Since the
ON/OFF signal is just only sent to switch, there is a low calculation time. Figure 4.7
indicates one transmission example with ON/OFF flow control.

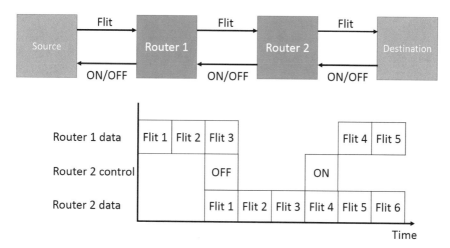

Fig. 4.7 ON/OFF flow control

4.2.3.2 Credit-Based Flow Control

In Credit-based flow control (CB), upstream nodes have information about the number of empty slots in downstream buffers. We call this information CN (Credit Number). Each time an upstream node sends a flit to downstream buffers, the number is decremented by one. When downstream buffers send some flits to other nodes, they also send a credit control signal to upstream routers, and when the upstream router receives the signal, the CN associated with the path is incremented appropriately. Figure 4.8 illustrates the data flow and an example of transmission. In this example, initially *Router 2* is blocked, and CN is decremented. Next *Router 2* starts sending flits and credit signals are emitted to *Router 1*, which receives the signal and restarts sending flits to *Router 2*.

4.2.3.3 ACK/NACK Flow Control

The above flow controls send signals from the downstream buffers to upstream ones and decide whether or not to send flits. On the other hand, ACK/NACK flow control [28] does not need to wait and calculate such signals from downstream buffers. In this flow control model, as flits are sent from source to destination, a copy is kept in each of the node buffers to resend it, if necessary, in case where some flits are dropped. An *ACK* signal is sent from a downstream node when a flit is received. When the upstream node receives this signal, it deletes its copy from its buffers. If the downstream node cannot or does not receive the correct flits, it sends *NACK* signal to the upstream node, and upstream node rewinds its output queue and starts resending a copy of the corrupted flit. Figure 4.9 depicts an example of this flow control.

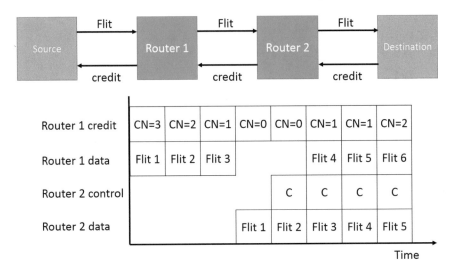

Fig. 4.8 Credit-based flow control

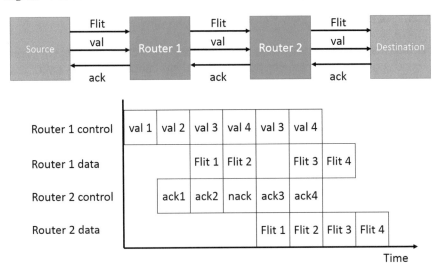

Fig. 4.9 ACK/NACK flow control

4.2.4 Routing Algorithms

This section will present some basic backgrounds and concept about routing algorithms. In general, the selected routing algorithm for a network is topology dependent. This section will give only a brief description about routing algorithms and their taxonomy. Routing algorithms can be classified according to several criteria:

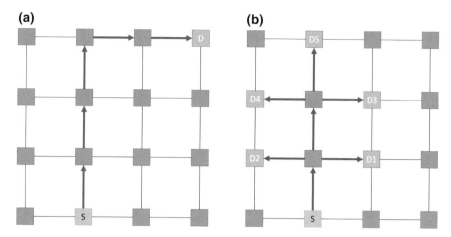

Fig. 4.10 Categorization of routing algorithms according to the number of destinations: **a** unicast, **b** multicast

- **Number of destinations:** According to the number of destination nodes, to which packets will be routed, routing algorithms can be classified into *unicast* routing and multicast routing as shown in Fig. 4.10. The unicast routing sends the packets from a single source node to single a destination node. The *multicast* routing sends the packets from a single node to multiple destination nodes. The multicast routing algorithm can be divided further into *Tree-based multicast routing* and *Path-based multicast* routing.
- **Routing Decision Locality:** According to the place where the routing decisions are made, routing algorithms (unicast or multicast routing) can be classified into *source routing* and *distributed routing*.
  As depicted in Fig. 4.11, in the distributed routing, there will be one header probe (for unicast routing case) containing the address of the destination node (probably also the source node). The routing information is locally computed each time the header probe enters a switch node. In the source routing, paths are computed at the source node. The precomputed routing information for every intermediate node, to where a message will travel, will be written in a routing probe. All routing probes that represent the routing paths from the source to destination node will then be assembled as packet headers for the message.
- **Adaptivity:** In all cases of the routing implementation seen so far, the routing algorithm can be either *deterministic* or *adaptive* (as represented in Fig. 4.12). In deterministic routing, the computed paths from a source and destination pair are statically computed and will always be similar. In adaptive routing algorithms, the paths from source to destination can be different, because the adaptive routing selects adaptively the alternative output ports. An output channel is selected based on the congestion information or the channel status of the alternative output ports. The adaptive routing algorithms generally guide messages away from congested or faulty regions in the network. Adaptive routing algorithms can be further clas-

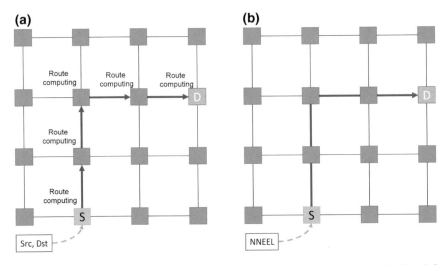

Fig. 4.11 Categorization of routing algorithms according to decision locality: **a** distributed, **b** source

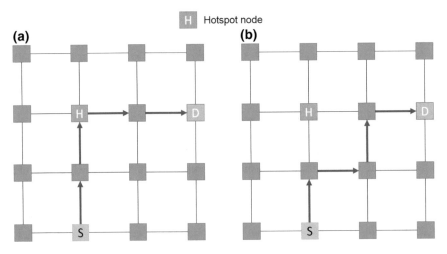

Fig. 4.12 Categorization of routing algorithms according to adaptivity: **a** deterministic, **b** adaptive

sified according to the number of alternative adaptive turns as *Fully adaptive* and *Partially adaptive* routing algorithms.

- **Minimality:** According to the minimality of the routing path, routing algorithms can be classified into *minimal* or *non-minimal* algorithm (see Fig. 4.13). The minimal adaptive routing algorithm will not allow a message to move away from its destination node. In other words, the message will always be routed closer to its destination node traversing the minimal number of hops to reach its destination. In the non-minimal algorithm which is also called as the detour routing algorithm,

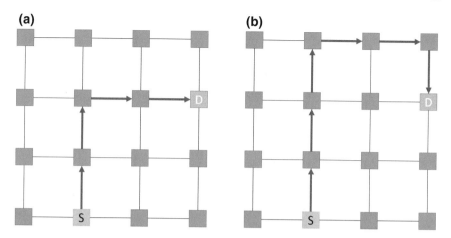

Fig. 4.13 Categorization of routing algorithms according to minimality: **a** minimal, **b** non-minimal

the message can be routed away from its destination node. This can be performed randomly or following some rules and restrictions usually found in adaptive routing [29].

4.2.4.1 Deadlock and Livelock Avoidance

Deadlock is caused by the cyclic dependency between packets in the network. It is one of the major issues in NoC systems which is caused when packets in different buffers are unable to progress because they are dependent on each other forming a dependency cycle. It can occur because packets are allowed to make all turns in clockwise and counterclockwise turn directions. Figure 4.14 illustrates a deadlock example in an adaptive NoC system. The dependency is caused by the flits exchange between R_{02} and R_{01}. Due to the presence of faults, the choices for a minimal routing is limited and both communications are dependent on each other; thus, none of them can make progress along the network. On the same figure, we can see that flits *Dest10* and *Dest00*, stored in the input-ports of R_{11} and R_{01} respectively, are victims of this deadlock; i.e., even their output-channel is free, they have to wait in the buffer until the blocking is resolved.

Virtual-Channel (VC) [25] is one of the most well-used techniques for deadlock avoidance. As illustrated in Fig. 4.15, VC divides the input-buffer in smaller queues which are independent of each other and managed by an arbiter. When a blockage happens in one VC, the other ones are not affected and they continue asking requests for their corresponding output-channels. In this fashion, non-blocked requests are served and their slots are freed to host other incoming flits.

Another technique used for deadlock-avoidance is called virtual-output-queue (VOQ) [30]. In VOQ, as shown in Fig. 4.16, the input-buffer is divided into different

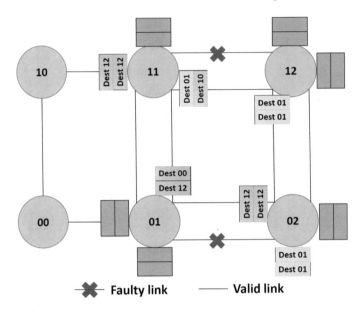

Fig. 4.14 Deadlock example in adaptive NoC systems

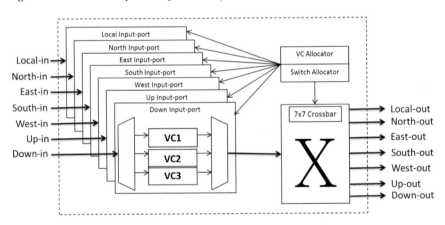

Fig. 4.15 Virtual-channel-based router architecture

queues to host incoming flits which are stored depending on their corresponding output-channel; i.e., VOQ (i,j) stores flits coming from input-port i wishing to access output-port j. For each output-channel, a 7×1 crossbar(i) is dedicated to handle the traversal of flits coming from the different input-channels and asking the grant for the output-channel(j).

Both VC and VOQ ensure deadlock-freedom; however, the employment of such techniques is costly in terms of hardware and implementation complexity. This is caused by the arbitration needed to handle the different requests coming from the

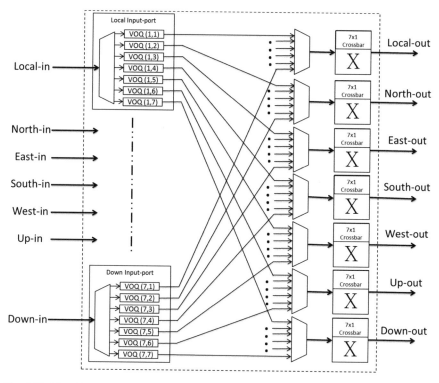

Fig. 4.16 Virtual-output-queue-based router architecture

multiple VCs/VOQs at each input-port. To solve this overhead, another solution for deadlock avoidance can be achieved by applying allowed turns and prohibiting one turn in every clock-wise and counter clock-wise turn direction. The prohibited turns will avoid cyclic dependency between packets in the network.

Some routing algorithms are solving the deadlock problem based on these prohibitions which are called *turn models*. The design of adaptive routing algorithms based on turn models has been introduced in [31]. The work has presented examples of turn models for adaptive routing algorithms in 2D mesh-based interconnection network.

If the packets are allowed to make non-minimal adaptive routing, then a problem called livelock configuration may occur. The livelock is a situation where a packet moves around a destination node but it never reaches the destination node. The livelock can be avoided by only allowing the packets to make minimal routing. However, if the non-minimal routing is allowed, then the mechanism to detect livelock must be implemented.

4.3 Hardware Design of On-Chip Network

After describing the architecture of the multicore on-chip network in the previous section, we now delicate this section to the actual hardware design of a mesh-based on-chip network using Verilog hardware description language. For simplicity, we only focus on the main building blocks of the on-chip network.

4.3.1 Topology Design

Figure 4.17 illustrates a mesh-based topology of a so called OASIS NoC (ONoC) and Fig. 4.18 illustrates the external connections to a router ($i = 0$ is Local port, $i = 1$ is "North" port, $i = 2$ is "East" port, $i = 3$ is "South" port and $i = 4$ is "West" port).

The Verilog RTL coding for a mesh topology design is shown in Listing 4.1. The parameters $X - WIDTH$ and $Y - WIDTH$ means network size, when $i == 1$, each router's north input port receives data from south port of $currentx, currenty + 1$ router. ($y - pos == Y - WIDTH - 1$) indicates the router position is north edge, so there are no input data from north port. All ports connections can be written using the same method.

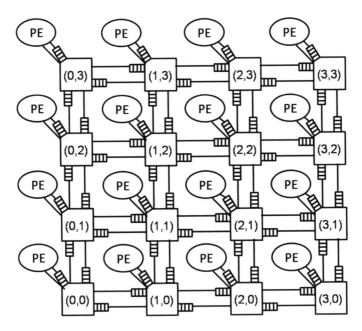

Fig. 4.17 4×4 mesh topology

Fig. 4.18 External connections to one router

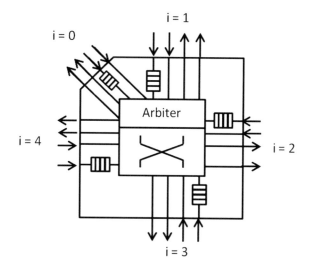

Listing 4.1 Verilog RTL coding for a mesh topology

```
1   //y loop
2   for (y_pos=0; y_pos<Y_WIDTH; y_pos=y_pos+1) begin:y_loop
3           //x loop
4       for (x_pos=0; x_pos<X_WIDTH; x_pos=x_pos+1) begin:x_loop
5
6       router #(NOUT, FIFO_DEPTH, FIFO_LOG2D, FIFO_FULL_LVL)
7           rtr(.clk(clk), .reset(reset),
8           .data_in(net_data_in[x_pos][y_pos]),
9           .data_out(net_data_out[x_pos][y_pos]),
10          .stop_in(net_stop_in[x_pos][y_pos]),
11          .stop_out(net_stop_out[x_pos][y_pos]),
12          .xaddr(x_pos['L2NET_SIZE-1:0]), .yaddr(y_pos['L2NET_SIZE
                -1:0]));
13
14  for (i=0; i<NOUT; i=i+1) begin:i0
15      //tile interface of router
16      if(i==0) begin
17        assign net_data_in[x_pos][y_pos]['WIDTH*(i+1)-1:'WIDTH*i] =
              data_in[('WIDTH*X_WIDTH
18          *y_pos)+('WIDTH*(x_pos+1))-1:('WIDTH*X_WIDTH*y_pos)+('
              WIDTH*x_pos)];
19        assign data_out[('WIDTH*X_WIDTH*y_pos)+('WIDTH*(x_pos+1))
              -1:('WIDTH*X_WIDTH*y_pos)
20          +('WIDTH*x_pos)] = net_data_out[x_pos][y_pos]['WIDTH*(i
              +1)-1:'WIDTH*i];
21        assign net_stop_in[x_pos][y_pos][i] = stop_in[(X_WIDTH*y_pos
              )+x_pos];
22        assign stop_out[(X_WIDTH*y_pos)+x_pos] = net_stop_out[x_pos
              ][y_pos][i];
23      end
24      //north edge of router
25      if(i==1) begin
26          if(y_pos==Y_WIDTH-1) begin
27              assign net_data_in [x_pos][y_pos]['WIDTH*(i+1)-1:'
                  WIDTH*i] = 0;
28              assign net_stop_in [x_pos][y_pos][i] = 1'b1;
29          end else begin
30              assign net_data_in [x_pos][y_pos]['WIDTH*(i+1)-1:'
                  WIDTH*i] = net_data_out[x_pos
```

```
31              ][y_pos+1]['WIDTH*(3+1)-1:'WIDTH*3];
32          assign net_stop_in [x_pos][y_pos][i] = net_stop_out[
              x_pos][y_pos+1] [3];
33        end
34      end
35      //east edge of router
36      if(i==2) begin
37        ...
```

4.3.2 Pipeline Design

ONoC architecture has three pipeline stages. The first stage mainly includes the *Buffer module*. The second stage includes the *Routing module, buffer overflow module*, and *scheduling module*. The last stage includes the *Crossbar module*.

Figure 4.19 illustrates the router micro-architecture. The five modules in the left side of the figure are the input port modules with buffers and routing modules. The other important module is the switch allocation (sw-alloc) module which mainly implements the scheduler and flow control modules. Finally, the crossbar module implements the crossbar circuitry and has an array of data input and output paths.

4.3.2.1 Input Port Design

A router in mesh-based NoC system has 5 input ports. Each input port has two main functions: buffering, and routing calculation. The *buffering* task design is shown in Listing 4.2. This module manages the FIFO pointers (lines 2–9 in Listing 4.2), and the *stop_out* signal (lines 11–22) for upstream router's flow control.

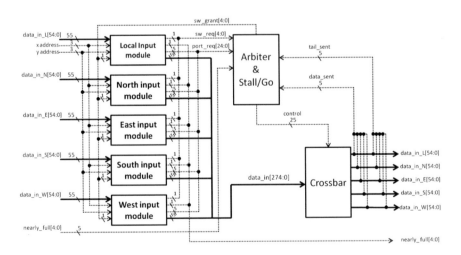

Fig. 4.19 ONoC router block diagram

The RTL code for the *Routing calculation* task is shown in Listing 4.3. As shown in the above code, a look-ahead XY routing method is performed. The next port address, which is used in routing calculation phase, is computed (lines 2–8 in Listing 4.3).

Listing 4.2 Verilog RTL coding for manging FIFO.

```verilog
always @(posedge clk) begin
    if (!reset) begin        //If out of reset
        if (enqueue) begin //Write a flit to the buffer
            fifo[tail_ptr] <= data_in;
            tail_ptr <= tail_ptr + 1;
        end
        if (dequeue) begin //Read a flit from the buffer
            head_ptr <= head_ptr + 1;
        end

        //nearly full signal = stop_out,
        if (((tail_ptr + FULL_LVL[LOG2D-1:0] + 1'b1)==head_ptr
            ) && enqueue && !dequeue)begin
            stop_out <= 1'b1;
        end
        if (((tail_ptr + FULL_LVL[LOG2D-1:0]) ==(head_ptr+1'b1)
            ) && !enqueue && dequeue)begin
            stop_out <= 1'b1;
        end
        if ((tail_ptr + FULL_LVL[LOG2D-1:0])==head_ptr)begin
            if ((enqueue && !dequeue) || (!enqueue && dequeue)
                )begin
                stop_out <= 1'b0;
            end
        end
    end
    else ...
end
```

Listing 4.3 Verilog coding for XY routing

```verilog
//assign next addresses
if (nextport == 'EAST) next_xaddr = xaddr + 1'b1;
    else if (nextport == 'WEST) next_xaddr = xaddr - 1'b1;
        else next_xaddr = xaddr;

if (nextport == 'NORTH) next_yaddr = yaddr + 1'b1;
    else if (nextport == 'SOUTH) next_yaddr = yaddr - 1'b1;
        else next_yaddr = yaddr;

//evaluate next port
if (next_xaddr == xdest) begin
    if (next_yaddr == ydest) route = 'SELF;
        else if(next_yaddr < ydest) route = 'NORTH;
            else route = 'SOUTH;
end else begin
    if (next_xaddr < xdest) route = 'EAST;
        else route = 'WEST;
end
```

4.3.2.2 Switch Allocator Design

The block diagram of the arbiter is shown in Fig. 4.20. Each row of the matrix means competitive inputs and has priority level. After the highest priority input is served, the

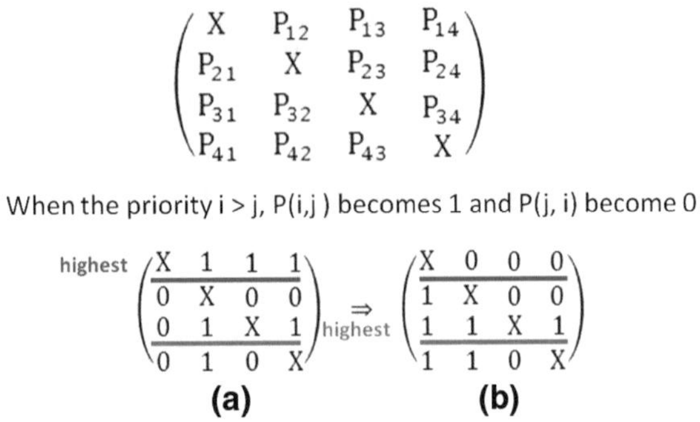

Fig. 4.20 Matrix arbitration example

$$\begin{pmatrix} X & P_{12} & P_{13} & P_{14} \\ P_{21} & X & P_{23} & P_{24} \\ P_{31} & P_{32} & X & P_{34} \\ P_{41} & P_{42} & P_{43} & X \end{pmatrix}$$

When the priority i > j, P(i,j) becomes 1 and P(j, i) become 0

$$\text{highest}\begin{pmatrix} X & 1 & 1 & 1 \\ 0 & X & 0 & 0 \\ 0 & 1 & X & 1 \\ 0 & 1 & 0 & X \end{pmatrix}_{\text{highest}} \Rightarrow \begin{pmatrix} X & 0 & 0 & 0 \\ 1 & X & 0 & 0 \\ 1 & 1 & X & 1 \\ 1 & 1 & 0 & X \end{pmatrix}$$

(a)   **(b)**

priority will be changed to lowest by inversing one's row and column. Figure 4.20a shows an example of how the matrix arbitration works.

The *Switch Allocator* includes a *Scheduler, Matrix Arbiter*, and a *Stop Go* flow control modules. Listing 4.4 shows part of the *Matrix Arbiter* module. Lines 1–15 generate grant i, and lines 17–23 calculate next state of all matrix elements. Finally, lines 25–30 update these states (Fig. 4.21).

Listing 4.4 Verilog coding for Matrix Arbiter

```
1    //Matrix Arbiter
2    generate
3        for (i=0; i<SIZE; i=i+1) begin:oll
4            for (j=0; j<SIZE; j=j+1) begin:ill
5                if (j==i)
6                    assign pri[i][j]=request[i];
7                else
8                if (j>i)
9                    assign pri[i][j]=!(request[j]&&state[j*SIZE+i]);
10               else
11               assign pri[i][j]=!(request[j]&&!state[i*SIZE+j]);
12           end
13                   assign grant[i]=&pri[i];
14       end
15   endgenerate
16
17   generate
18   for (i=0; i<SIZE; i=i+1) begin:ol2
19       for (j=0; j<SIZE; j=j+1) begin:il2
20           assign new_state[j*SIZE+i]=(success&&((state[j*SIZE+i
                    ]&&!grant[j])||(grant[i])))
21                   ||(!success&&state[j*SIZE+i]);
22       end
23   end
24   endgenerate
25
26   always@(posedge clk) begin
27       if (reset) state<=-1;
28       else begin
29       if (|request) state<=new_state;
30       end
31   end
```

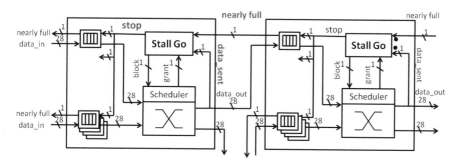

Fig. 4.21 Stall-go block diagram

For the flow control, ONoC employs *stop-go* scheme. This technique ovoids buffers overflow. Data transfer is controlled by signals indicating the buffers condition. In the absence of *stall-go* function, the receiver cores need to judge whether there are dropped packets or not. If so, the transmitter must resend the dropped packets using a receiving request signal from master cores. In addition, *stall-go* scheme reduces blocking, but at the same time it may increase latency in some situations. Figure 4.22a illustrates the state machine of this approach and Fig. 4.22b shows the input FIFO state of *nearly full* signal output.

Code 4.5 shows the RTL code of the state machine. State *Go* indicates that the receiving FIFO can store more than two flits. State *Sent* means that it can store one flit and state *Stop* means that it cannot store any more flits. Figure 4.21 shows the *stop-go* control flow scheme.

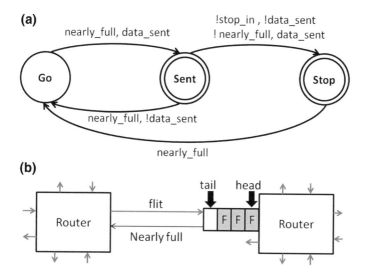

Fig. 4.22 **a** State machine design, **b** Nearly full signal output

Fig. 4.23 Arbiter control signals

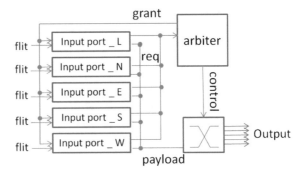

Listing 4.5 Verilog coding for the stall-go state machine

```
1   //Stall-go state machine.
2      always @ (posedge clk) begin
3
4      if (!reset) begin
5          if ((state=='GO) && stop_in && data_sent)
6              state <= 'SENT1;
7          if (state=='SENT1) begin
8              if (stop_in && !data_sent)
9                  state <= 'GO;
10             if (!stop_in && data_sent)
11                 state <= 'STOP;
12         end
13
14     if ((state=='STOP) && stop_in) // stop_in = nearly_full
15         state <= 'GO;
16     end else
17         state <= 'GO;
18     end
19
20     assign blocked = ( ((state=='STOP) && !stop_in) || ((state=='
       SENT1) && !stop_in &&
21         data_sent) );
```

4.3.3   Crossbar Design

The crossbar in ONoC architecture is a an important module that connects multiple inputs to multiple outputs. It is basically an assembly of single switches between multiple inputs and multiple outputs. Code 4.6 shows the Verilog RTL code for the crossbar (Fig. 4.23).

Listing 4.6 Verilog RTL code for the crossbar

```
1   // Crossbar
2   generate
3       for (i=0;i<NOUT;i=i+1) begin:output_loop
4       mux_out #(NIN, WIDTH) cbar_mux(.cntrl(cntrl_reg[NIN*(i+1)
            -1:NIN*i]),
5   .data_in(data_in), .data_out(data_out[WIDTH*(i+1)-1:WIDTH*i]));
6       end
7   endgenerate
8
9   //mux_out
10  generate
11      //loop over each bit of data
12      for (i=0;i<WIDTH;i=i+1) begin:bit_loop
13          assign data_out[i] = mux(cntrl, data_bits[i]);
14          //loop over each input channel
15          for (j=0;j<n_in;j=j+1) begin:input_loop
16              assign data_bits[i][j] = data_in[WIDTH*j+i];
17          end
18      end
19  endgenerate
20
21  function mux;
22      input [n_in-1:0] cntrl;
23      input [n_in-1:0] data_in;
24      integer i;
25
26  begin
27      mux = 0;
28      for (i=0; i<n_in; i=i+1) begin
29          if(cntrl[i] == 1'b1) mux = data_in[i];
30      end
31  end
32  endfunction // mux
```

4.3.4 Limitations of Regular Mesh Topology

The ONoC communication architecture considered so far is based on regular mesh topology. This provides well-controlled electrical parameters and reduced power consumption across the links. However, because of the nonexistence of short (fast) paths between remotely situated nodes, such architectures may suffer from long packet latencies. To solve this problem, a so called *short-path-link (SPL)* is used [32] to shorten the path; thus, decreasing the latency.

Figure 4.24 shows a simple example showing SPL insertion between two remote nodes. To support optimization with SPL insertion, a new port should be added to the five ports in each router. Consequently, each router will have six ports instead of only five ports in fully regular mesh topology.

Fig. 4.24 Short-path-link
(SPL) insertion example

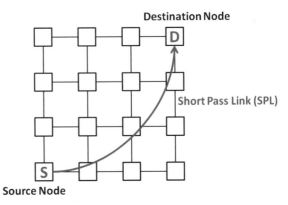

4.3.5   SPL Insertion Algorithm

The SPL algorithm selects communication paths that need optimization. The available SPL resources should be decided first so that the power and area are not increased. Then, the communication costs for all communication patters are calculated using the communication frequencies and the distance between different nodes. Depending on the output of this computation, the SPL is inserted to the highest communication cost. After adding an SPL, the algorithm loops until the available SPL budget is exhausted.

$$f_{ij} = \max \frac{V_{ij}}{\sum_p \sum_{p \neq q} V_{pq}} \tag{4.1}$$

$$d_M(i, j) = |i_x - j_x| + |i_y - j_y| \tag{4.2}$$

$$C_{ij} = f_{ij} \times d_M(i, j) \tag{4.3}$$

where S is the available resource, i is the target sender router, j is the target receiver router, f_{ij} is the target communication frequency, and C_{ij} is the target communication total cost.

Figure 4.25 shows the SPL insertion algorithm. Equation 4.1 is used to measure the communication frequency by calculating the whole communication with all neighbor nodes and the target communication for the whole neighbor nodes usability volume.

The whole communication volume is expressed by $\sum_p \sum_{p \neq q} V_{pq}$; where p indicates the sender node, q indicates the receiver node. Notice that p and q are always neighbors. Then, the target communication frequency is expressed by V_{ij}; where i indicates the sender node, and j indicates the receiver node.

To calculate the distance (number of hops) of communications, *Manhattan distance* is employed (Eq. 4.2). The address of i node is expressed by (i_x, i_y), and the address of j node is expressed by (j_x, j_y). Finally, the total cost calculation is computed using Eq. 4.3 with computed values from the previous two equations.

Fig. 4.25 SPL insertion
algorithm

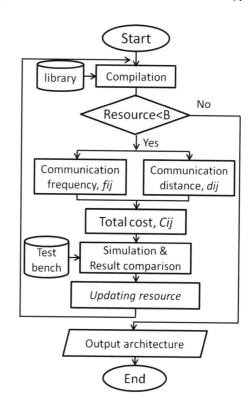

4.3.5.1   SPL Complexity

Tables 4.1 and 4.2 show the area utilization of 5-port and 6-port routers respectively.
From these results, we can see that the 6-port router's ALUTs utilization is increased
by 33.2% and the registers utilization is increased by 56.6% when compared with
5-port router.

Table 4.1 Area utilization for a 5-ports router

Parameters		Input port	Switch allocator	Crossbar	Total
ALUTs	1-port	71(7.4%)	300(31%)	310(32.1%)	965
	5-ports	355(36.8%)			
Registers	1-port	72(15.2%)	90(18.9%)	25(5.3%)	475
	5-ports	360(75.8%)			

Table 4.2 Area utilization for 6-port router

Parameters		Input port	Switch allocator	Crossbar	Total
ALUTs	1-port	75(5.8%)	469(36.5%)	366(28.5%)	1285
	6-ports	450(35%)			
Registers	1-port	99(13.3%)	144(19.4%)	36(4.8%)	744
	6-ports	594(79.8%)			

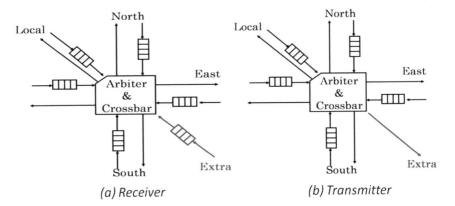

(a) Receiver   (b) Transmitter

Fig. 4.26 Extra-port insertion

4.3.5.2 Hardware Modification for SPL Support

Initially, each router has 5 in/out ports: Local, North, East, South, and West. In order to optimize the system with SPL approach, it is essential to add another port. Figure 4.26 shows the extra-port addition for both sender and receiver nodes. Flit structure also needs to be modified to support SPL. Initially, ONoC has 5 bits dedicated used for the *next-port* field direction. To support SPL, it is also necessary to modify the network connection between routers and also extend the *next-port* field by 1 bit. That is, from 5 to 6 bits. Figures 4.27, 4.28 and 4.29 illustrate the application mapping with SPL links. We assume here that the resource budget is 5% of the original area utilization. Code 4.7 shows the modified code for SPLs insertion.

We mainly modified the loop function for the mesh topology because it is necessary to simplify the connection of all routers (lines 1–4). After that, designers can easily insert one or more SPLs. For example, lines 15–17 show the address (0,3) north output connects to (1,0) south input port, and the address (1,0) south port has no connections. Thus, the south port can use SPL without adding an extra port.

As we mentioned, ONoC router employs *look-ahead* XY routing, so the routing stage calculates next router's output direction. The SPL is inserted from source node to destination node directly. Lines 6–13 in code 4.8 shows routing calculation for (0,3) to (1,0) communications.

Fig. 4.27 Dimension reversal with 2 SPLs

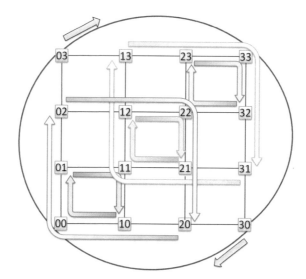

Fig. 4.28 Hotspot with 2 SPL

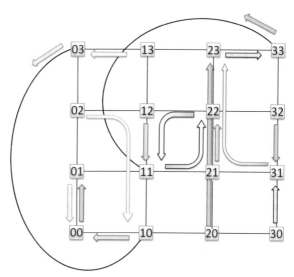

Listing 4.7 Code Modification for NoC architecture

```
1   // code Modification for SPL.
2   //y loop
3   for (y_pos=0; y_pos<Y_WIDTH; y_pos=y_pos+1) begin:y_loop2
4       //x loop
5       for (x_pos=0; x_pos<X_WIDTH; x_pos=x_pos+1) begin:x_loop2
6           ...
7           ///(1,0)
8           if (x_pos == 1 )begin
9               if (y_pos == 0)begin
10                  //tile interface of router, x_pos = 1, y_pos = 0,
                        i = 0
11                  ....
```

```
12      //north edge of router, x_pos = 1, y_pos = 0, i =
                1
13      . . . .
14      //east edge of router, x_pos = 1, y_pos = 0, i = 2
15      . . . .
16      //south edge of router, x_pos = 1, y_pos = 0, i =
                3
17      assign net_data_in [x_pos][y_pos]['WIDTH*(3+1)-1:'
                WIDTH*3] = net_data_out [0] [3] ['
18              WIDTH*(1+1)-1:'WIDTH*1];
19      assign net_stop_in [x_pos][y_pos] [3] =
                net_stop_out [0] [3] [1];
20
21      //west edge of router, x_pos = 1, y_pos = 0, i = 4
22      . . .
23  end
24  . . .
```

Listing 4.8 Code Modification for the look-ahead routing

```
1   if(nextport == 'WEST) begin
2       next_xaddr = xaddr - 1'b1;
3   end else if (nextport == 'EAST) next_xaddr = xaddr + 1'b1;
4   else next_xaddr = xaddr;
5
6   if(nextport == 'NORTH)begin
7       if((xaddr==0&&yaddr==3)&&(xdest==1&&ydest==0))begin
8           next_xaddr = 1;
9           next_yaddr = 0;
10      end
11      else next_yaddr = yaddr + 1'b1;
12  end else if (nextport == 'SOUTH)    next_yaddr = yaddr - 1'b1;
13  else next_yaddr = yaddr;
```

4.3.6 Network Interface Design

In Network-on-Chip architectures, the network interface (NI) plays an important role of acting as interface between IP cores and the communication infrastructure.

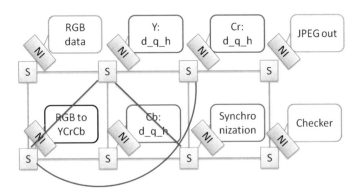

Fig. 4.29 JPEG encoder with 3 SPL

In general, a NI includes a front-end and a back-end submodules. The front-end module implements the communication protocol adopted by the core and the back-end module is in charge of implementing basic communication services, such as packetization/depacketization, control flow, and routing-related functions.

The NI must provide low area overhead because NoC designs are generally constrained by area and power. In addition, a good NI design must provide throughput and/or latency guarantees, which are essential for the design of NoC-based complex multicore SoCs.

Whether it is used in on-chip network on off-chip network, the NI's main job is to convert messages to packets and packets to messages. In NoC architecture, a core is connected to router through the NI and it communicates within the network using packets. Design of the NI needs to consider the I/O structure of the core and the protocols used in the NoC at physical, data link and network layers.

The NI functionality can be divided into two parts: the *Core part*, and the *Network part* as illustrated in Fig. 4.30. The *Network part* handles interface to the router; wile the *Core part* is connected with core and it deals with the data and address bus width, and control signals.

There are two main types of NIs: (1) Network interface for source routing, and (2)Network interface for distributed routing. This chapter only focuses on the design of the distributed routing NI type.

Source Routing Network Interface: As the name indicates, in source routing the information about packet route is embedded in the packet's header at the source end. In this way, the source node makes all routing decisions before the packet is transmitted into the network (NoC). The NI contains a routing table filled with routing information. The sender's NI selects route path from its table and places this information in the packet header. Then, packet is transmitted in the network through the NI. When a given packet reaches an intermediate router, the route path is read from the packet's header and forwarded to the corresponding neighbor router until it reaches its destination.

Fig. 4.30 Nigh-level view of the network interface

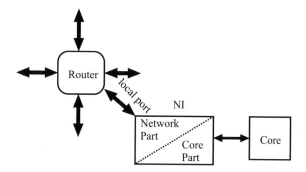

Distributed Routing Network Interface: In distributed routing NI, a destination address is added in the packet's header. Unlike source routing NI, it does not have a path information table. So, the circuit size is relatively smaller than the source routing NI's one.

In distributed routing protocol, the routing functions are implemented in each NoC router. The header, which is generally compact, carries the destination address and some control bits. In this way, each router contains information about the neighbor routers. When the packet arrives at the input port of the downstream router, the route path is selected either by looking up the routing table or calculating the routing path in hardware.

The advantage of the distributed routing is that it can be easily expanded to support adaptive routing. The disadvantage is the large additional hardware for execution of routing logic, and the extra memory unit used to store routing tables. Distributed routing is suitable for regular topologies, such as mesh topology.

4.3.6.1 Design Decisions of Distributed Routing NI

The block diagram of the designed NI is given in Fig. 4.31. FPGA and Quartus II software design tools [33] were used for the prototyping of this interface. The used core is a Nios II processor [34], which is a configurable 32-bit RISC soft core processor.

As shown in Fig. 4.31, the NI has different internal blocks, including buffers, flitizer, deflitizer, and controllers. The controller is the main module of the NI and it controls packet transmission from core to router and from router to core. When a core wants to send a packet to another core, it first stores the packet in the buffer of the NI. When the router is ready to receive the packet, the NI converts the packet into flits and sends the flits to the router. Similarly, when the NI receives flit(s) from the router, the NI converts them/it into a packet and stores it/them in the buffer. Then when the core is ready to receive the packet, the NI transfers the packet to the core. Some control signals are used for the communication between core and NI and NI and router. A wormhole switching technique is used in the packet transmission from NI to router and from router to NI.

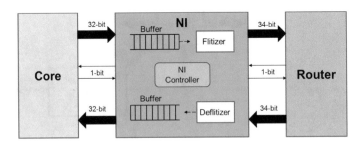

Fig. 4.31 Distributed routing NI architecture block diagram

Before we start talking about the actual design of the NI, we need first to make several design decisions. We mainly need to decide: (1) network size, (2) packet size, (3) buffer size, (4) communication Protocol, (5) packet buffering, and (6) packet/flit format.

Network Size Decision: Network (NoC) size is a very important decision which we need to make. The network size depends mainly on the target application and on how much parallelize we have. In other words, after mapping the application (task-graph) to the NoC architecture (refer to Chap. 3), we are able to know the number of needed cores. If, for example, after several simulations and profiling, we found that we need 62 cores to run a given application, the network size should be, then, 8×8 (64 cores). Notice that with this size, there will be unused routers since we have only 62 cores (1 router for each core).

For our network interface, we assumed the NoC size of 8×8. Thus, 6-bit are needed to represent one destination address direction. Since we have two directions (X-Y coordinates) we need 12-bit for the complete address. The 12-bit will be embedded in the header of the packet.

Packet Size Decision: Packets in a given NoC system can be of different sizes. The size depends on the application, target platform, and available hardware resources. Therefore, we need to decide the packet size so that we can decide the maximum buffer size. This is also very important because NoC design is area and power constrained.

In this design and in order to keep the design simple, we assume that the maximum size of the packet will be 512-bit, i.e., 16×32-bit flits. In distributed routing, a packet can have, then, 1 flit minimum and 16 flit maximum.

Buffer Size Decision: The role of a buffer in the NI is to temporarily store the packets while they are transferred from the source core to the destination core. The size of the buffer in the NI should be equal or larger than the packet size. The idea is to have the maximum size of the buffer at least equal to the maximum size of a packet. Since our packet size is fixed to 512-bit, the buffer size is also 512-bit.

Communication Protocol and Flow Control Decisions: We used *Ready-to-Receive (RTR)*-based scheme as a communication protocol between core and NI and between NI and router. In this scheme, two 1-bit signals and 1 WR signal are used for handshaking signals. We assume that *phit* size is the same as *flit* size.

This NI design will be tested with Altera Nios II core which can be connected with various external peripherals. Nios II support 32-bit PIO width. Thus, it can send/receive 32-bits of data at a time.

Packet Format Decision: As we mentioned, the maximum size of a packet is fixed to 512-bits. The packet is divided into three parts: HEADER, BODY, and END. The packet's HEADER contains the first 32-bits of the packet. The last 32-bits is the END of the packet and the remaining bits of the packet are reserved for the payload (BODY). The packet format for the distributed routing NI is shown in Fig. 4.32. The size of the packet's HEADER is 32-bits. Since the maximum size of the NoC is

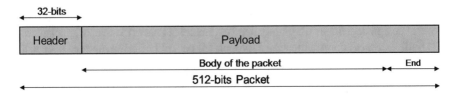

Fig. 4.32 Packet format

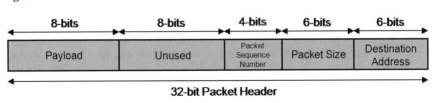

Fig. 4.33 Packet HEADER format

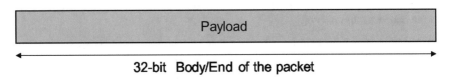

Fig. 4.34 BODY and END format

8×8, a minimum 6-bits are required to represent the node address in the network. In the HEADER, the first 6-bits represent the *Destination Address* of the core in the network. The next 6-bits represent the *Packet Size*, which helps tracking the arrival of the whole packet. The next 4-bits carry the *Packet-Sequence-Number*. This number is used to rearrange the packet in the correct order at the destination core. The next 8-bits are *Unused* and are reserved for future extension. The remaining 8-bits are for *Payload* data field. The HEADER format is shown in Fig. 4.33.

The formats of the BODY and END flits are shown in Fig. 4.34.

Flit-Level Decision: After receiving a packet from the core, the NI converts the packet into flits. This process is called *Flitization*. A packet can have minimum 1 flit and maximum 16 flits. The size of a flit is kept fixed and is equal to 34-bits. The first 2-bits of each flit indicate *Flit Type*. Each type of flit is encoded as shown in Table 4.3.

Table 4.3 Flit Types and Coding

Flit type	Code
Single Flit with full Payload	00
HEADER flit	01
Body flit	10
End flit	11

Flit Format after Flitization: The HEADER flit is the first flit of a packet that enters into the network through the NI. In distributed routing, this flit carries first 24-bits as control information and next 2-bits are unused while the rest 8-bits are payload. HEADER flit is used for locking the path for the following body flits and an end flit while traversing through the network.

Two bits are used to decode the type of HEADER flits. Code 00 is used when the original packet from the core is only 32-bits including the packet header. In this case, there will be only 1 flit that corresponds to the original packet and there will be no BODY and END flits.

When the code is 01, this means that the original packet is more than 32-bits. In this case, the packet can have both BODY and END flits or just an END flit. The HEADER flit format is shown in Fig. 4.35. *BODY Flit*: The BODY flit always follows the HEADER flit and carries the payload. After flitization, a packet may have a minimum of 0 BODY flit and maximum of 14 BODY flits, depending on the payload size in the original packet. The BODY flit is represented by code 10 and its format is shown in Fig. 4.36. *END Flit*: The END flit is the last flit in the group flits corresponding to a particular packet. It follows the last BODY flit. It unlocks the path for the packet to which it belongs. It should be noted here that the path was locked by the HEADER flit of the same group of flits. The END flit format is shown in Fig. 4.37.

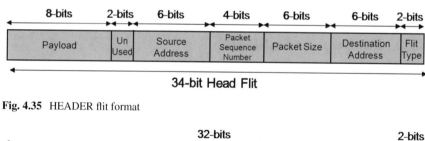

Fig. 4.35 HEADER flit format

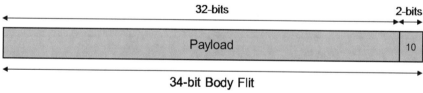

Fig. 4.36 BODY flit format

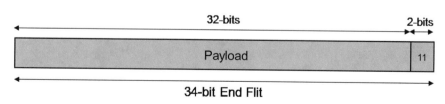

Fig. 4.37 END flit format

Table 4.4 Summary of decisions for distributed routing NI

Maximum NoC size	Maximum packet size (Bits)	Buffer size (Bits)		Flit size (Bits)
		Buffer 1	Buffer 2	
8 × 8	512	512	512	34

Flit Format After Deflitization: The process of converting the flits into a packet is called *Deflitization*. The *Deflitization* process starts after receiving the 34-bits HEADER flit from a router and continues until the END flit is received. *Deflitization* is needed for all cores in the network.

HEADER Flit: When the NI receives the 34-bits HEADER flit from the router, it removes the *Flit Type* and the *Destination Address* bits from the above flit. After that, the *Source Address* bits are shifted to the most right position. The *Packet Size* and *Packet Sequence Number* bits are also shifted to LSB (Least Significant Bit) side by 2-bits. The next 8-bits are unused and the remaining 8-bits are payload. The new created 32-bits packet HEADER (see Fig. 4.38) is stored in the NI buffer.

Both BODY and END flits are deflitized by removing the *Flit Type* bits and the rest 32-bits payload is transferred to the buffer in the NI. The formats of both BODY and END flits after deflitization are the same and shown in Fig. 4.39.

Summary of Design Decisions: The design decisions at all levels for the distributed routing NI are shown in Table 4.4.

4.3.6.2   Distributed Routing NI Design

The detailed internal structure of the NI for distributed routing is shown in Fig. 4.40. It consists of 6 internal blocks: C2R-Buffer, Flitizer, C2R-Controller, R2C-Buffer, Deflitizer, and R2C-Controller. Each block performs its defined specific job.

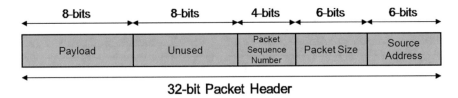

32-bit Packet Header

Fig. 4.38 Format of packet header after deflitization

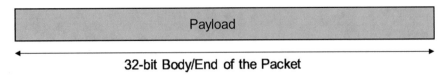

32-bit Body/End of the Packet

Fig. 4.39 Format of BODY/END flits after deflitization

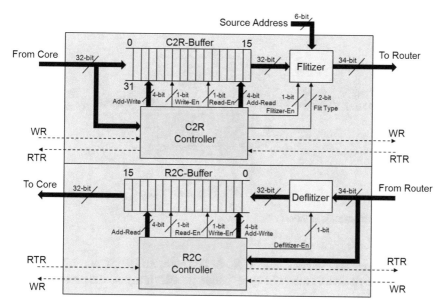

Fig. 4.40 Internal structure of NI for distributed routing

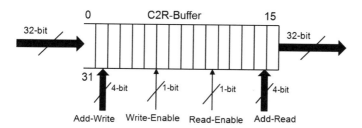

Fig. 4.41 C2R-buffer

The NI has different blocks and control signals as shown in Fig. 4.40.

Core-to-Router (C2R) Buffer: C2R-Buffer is a FIFO structure which is connected to the input port of the NI from the core side. The C2R buffer has 16×32 entries. Whenever the above buffer receives the "Write-Enable" signal from the C2R-Controller, it stores a packet coming from the core at a particular location specified by the "Add-Write" signal from C2R-Controller. Similarly, whenever it receives the "Read-Enable" signal from the C2R-Controller, it sends the chunk from the address location which is specified by the "Add-Read" signal (Fig. 4.41).

Flitizer Module Architecture: As we mentioned earlier, the process of converting a packet into flits is called *flitization*. The input and output signals to the flitizer module are illustrated in Fig. 4.42. When the "Flitizer-Enable" signal arrives from the C2R-Controller, the flitizer module starts working on the flitization process; it reads the 32-bits of a packet from the C2R-Buffer. If the "Flit Type" value is "00,"

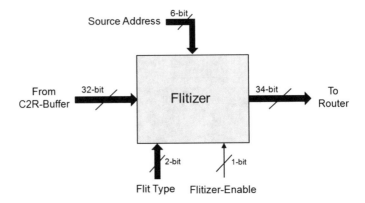

Fig. 4.42 Flitizer module architecture

it means the packet contains only 1 flit. In this case, no BODY and END flits are present in the packet. If the "Flit Type" value is "01," it means the packet contains more than 1 flit.

The flitizer circuit adds 2-bits flit type in the "Flit Type" field and 6-bits source address in the packet header, i.e., from bit numbers 18–23, and creates a 34-bits HEADER flit. When it receives the "Flit Type" signal ("10" or "11"), it assumes that the incoming packet from the C2R-Buffer is BODY or END of the packet respectively. In this situation, the flitizer just adds the "Flit Type" to the flit at the field, creates a 34-bits BODY or END flits. After the flitization process completes, flits are transferred to the router.

Core-to-Router (C2R) Controller: The Core-to-Router (C2R) is also a very important block in the NI since it generates several important control signals. The C2R controller consists of several modules as shown in Fig. 4.43. The C1 counter is a 6-bits counter and is used to count the total number of payload bytes of the packet coming from a given core to the NI. Initially, C1 is set to "000000." When a packet header arrives from the core, the corresponding bits in the packet header, which represents the size of payload bytes, will be stored in this counter.

The C2 counter is 4-bits counter and is used to locate the address of C2R-Buffer to store the received packet from the core. Initially, its value is also set to "0000." The C2 value is incremented by 1 whenever a new chunk of the packet is stored in the C2R-Buffer.

The C3 counter counts the total number of payload bytes (packet size) that has been transferred to router from C2R-Buffer. Initially, its value is also set to "000000." Similar to C1 counter, when a packet header is received from the core, the corresponding bits in the packet header will be stored in C3 counter.

The C4 counter is used to locate the address of the C2R-Buffer from where the chunk of the packet has to be transferred to flitizer. Whenever a chunk of the packet is sent from C2R-Buffer to flitizer, the counter value will be incremented by 1.

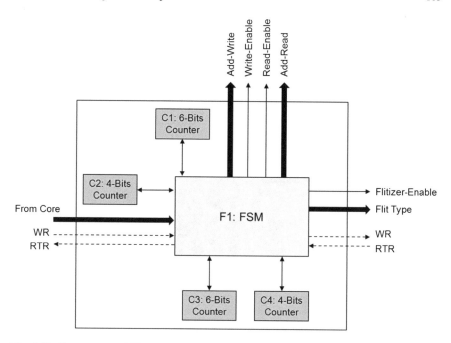

Fig. 4.43 Core-to-router (C2R) controller architecture

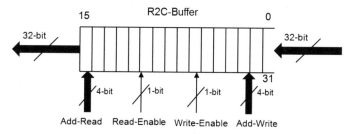

Fig. 4.44 Router-to-core (R2C) buffer

Router-to-Core (R2C) Buffer: The Router-to-Core (R2C) has a 16 entries FIFO buffer connected to the output port of the NI (see Fig. 4.44). Whenever it receives the "Write Enable" signal (high state) from R2C-Controller, it stores the flit (coming from deflitizer) at a specified address location. The address location is specified by the "Add-Write" signal from R2C-Controller. Similarly, whenever it receives the "Read-Enable" signal, it sends the stored flit from the specified address location of R2C-Buffer to the core (Nios II core in our case).

Deflitizer Module Architecture: The deflitization process starts whenever Deflitizer receives the "Deflitizer-Enable" signal from the R2C-Controller and then it reads a 34-bits flit from a router's port (see Fig. 4.45). It should first check the "Flit Type" bits. If it is "00" or "01," the Deflitizer simply removes the "Flit Type" and "Destination

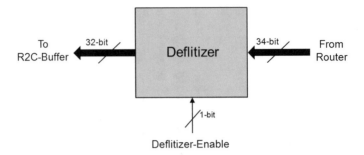

Fig. 4.45 Deflitizer module architecture

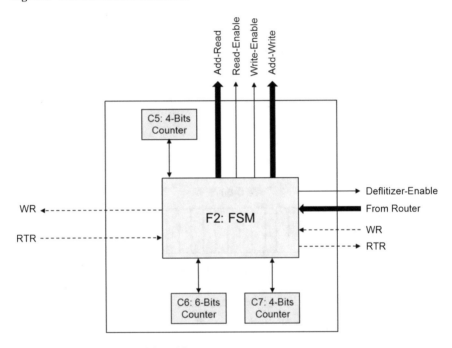

Fig. 4.46 R2C controller module architecture

Address" bits from the flit and shifts the "Source Address" bits to the "Destination Address" field and creates a 32-bits packet header. The created packet HEADER should exactly match the one that was created at the source. We have to note that only "Destination Address" bits are replaced by "Source Address" bits and the rest of the header bits remains the same. As soon as the deflitization process is completed, the created packet HEADER will be sent to R2C-Buffer.

Router-to-Core (R2C) Controller: The router-to-core (R2C) module is responsible for controlling the communications from the router to the core. This module consists of different components, including a finite-state machine (FSM) component. The block diagram of the R2C module is shown in Fig. 4.46.

4.4 Chapter Summary

The interconnection structure supporting future complex multi and many core SoCs will be closer to a sophisticated network than to current bus-based solutions. Such network must provide high throughput and low latency while keeping area and power consumption low. NoCs provide a good way of realizing interconnections on silicon and largely alleviate the limitations of bus-based solutions. NoC is a scalable interconnect with a huge potential to handle the increasing complexity of current and future multicore SoCs. In such paradigm, cores are connected via a packet-switching communication network on a single chip.

This chapter presented in details architecture and design of a real Network-on-Chip, which utilizes a Short-Path-Link (SPL) insertion customization to reduce the communication latency which directly affects the overall system performance.

References

1. A. Ben Abdallah, M. Sowa, basic network-on-chip interconnection for future gigascale mcsocs applications: communication and computation orthogonalization, in *Proceedings of Tunisia-Japan Symposium on Society, Science and Technology (TJASSST)* 4–9th Dec 2006
2. W.J. Dally et. al., Route packets, not wires: on-chip interconnection networks, in *the Proceedings of DAC* (2001), pp. 684–689
3. F.G. Morales et al., HERMES: an infrastructure for low area overhead packet-switching networks on chip. Integr. VLSI J. **38–1**, 69–93 (2004)
4. F.A. Samman, T. Hollstein, M. Glesner, Multicast parallel pipeline router architecture for network-on-chip, in *Proceedings of the Conference on Design, Automation and Test in Europe (DATE 08)* (Munich, Germany, 2008), pp. 1396–1401
5. A. Ben Abdallah, M. Nakamura, A.B. Ahmed, M. Meyer, Y. Okuyama, Fault-tolerant router for highly-reliable many-core 3D-NoC systems, in *Proceedings of the 3rd International Scientific Conference on Engineering and Applied Sciences (ISCEAS 2015)*, (Okinawa, Japan, 29–31 July 2015)
6. A.B. Ahmed, A. Ben, Abdallah, Adaptive fault-tolerant architecture and routing algorithm for reliable many-core 3d-noc systems. J. Distrib. Comput. **9394**, 30–43 (2016)
7. A.B. Ahmed, A. Ben, Abdallah, Graceful deadlock-free fault-tolerant routing algorithm for 3D network-on-chip architectures. J. Parallel Distrib. Comput. **74**(4), 2229–2240 (2014)
8. K.N. Dang, M. Meyer, Y. Okuyama, A. Ben Abdallah, X.-T. Tran, A Soft-error resilient 3D network-on-chip router, in *Proceedings of the IEEE 7th International Conference on Awareness Science and Technology (iCAST 2015)* (22–24 Sept. 2015)
9. K.N. Dang, Y. Okuyama, A. Ben Abdallah, Soft-error resilient network-on-chip for safety-critical applications, in *Proceedings of the IEEE International Conference on Integrated Circuit Design and Technology (ICICDT)* (27–29 June 2016)
10. S. Kumar, A. Jantsch, J.-P. Soininen, M. Forsell, et al., A network on chip architecture and design methodology, VLSI, 2002, in *Proceedings of IEEE Computer Society Annual Symposium* (25–26 April 2002), pp. 105–112
11. P.P. Pande, C. Grecu, M. Jones, A. Ivanov, R. Saleh, Performance evaluation and design trade-offs for network-on-chip interconnect architectures. IEEE Trans. Comput. **54**(8), 1025–1040 (2005)
12. E. Salminen, A. Kulmala, T. Hamalainen, On network-on-chip comparison, in *Euromicro DSD* (2007), pp. 503–510

13. A.V. de Mello, L.C.O.F.G. Morales, N.L.V. Calazans, *Evaluation of Routing Algorithms on Mesh Based NoCs* (Technical report, FACULDADE DE INFORMATICA - PUCRS, Brazil, 2004)
14. M. Li, Q.A. Zeng, W.-B. Jone, DyXY - a proximity congestion-aware deadlock-free dynamic routing method for network on chip, in *Proceedings of Design Automation Conference* (2006), pp. 849–852
15. E. Bolotin, I. Cidon, R. Ginosaur, A.N.D.A. Kolodny, *QNoC: QoS architecture and design process for network-on-chip* (J. Syst, Arch, 2004)
16. L. Bononi, N. Concer, Simulation and Analysis of Network on Chip Architectures: Ring, Spidergon and 2D Mesh. DATE (2006), pp. 154–159
17. J.S.K. (Ed.), *Parallel MIMD Computation: HEP Supercomputer and Its Applications* (MIT Press, Cambridge, MA, 1985)
18. R.S. Arvind, *Nikhil, Executing a Program on the MIT Tagged Token Dataflow Architecture*, Lecture Notes in Computer Science (Springer, Berlin/Heidelberg, 1987), p. 129
19. J. Gurd, C.C. Kirkham, I. Watson, The manchester prototype dataflow computer. Commun. ACM **28**(1), 3452 (1985)
20. S. Kumar, A. Jantsch, J.-K. Soininen, M. Forsell, M. Millberg, J.O Berg, K. Tiensyrja, A. Hemani, A network on chip architecture and design methodology, in *Proceedings of IEEE Computer Society Annual Symposium on VLSI* (2002), pp. 105–112
21. P. Martin, Design of a virtual component neutral network-on-chip transaction layer, in *Proceedings of Design, Automation and Test in Europe Conference and Exhibition (DATE05)* (2005), pp. 336–337
22. I. Saastamoinen, D.S.-Tortosa, J. Nurmi, Interconnect IP node for future systemon- chip designs, in *Proceedings of the 1st IEEE International Workshop on Electronic Design, Test and applications (DELTA02)* (2002), pp. 116–120
23. M. Sgroi, M. Sheets, K. Keutzer, S. Malik, J. Rabaey, A.S. Vincentelli, Addressing the system-on-a-Chip interconnectwoes through communication-based design, in *Proceedings of the 38th Design Automation Conf. (DAC01)* (2001), pp. 667–672
24. D.S. Tortosa, T. Ahonen, J. Nurmi, Issues in the development of a practical NoC: the Proteo Concept. Integr. VLSI J. Elsevier **38**(1), 95–105 (2004)
25. W.J. Dally, Virtual-channel flow control. IEEE Trans. Parallel Distrib. Syst. **3**(2), 194–205 (1992)
26. W.J. Dally, C.L. SEITZ. The torus routing chip. J Distrib. Comput. **1**(3), 187–196 (1986)
27. W.J. Dally, Performance analysis of k-ary n-cube interconnection networks. IEEE Trans. Comput. **C–39**(6), 775785 (1990)
28. A. Pullini , F. Angiolini , D. Bertozzi, L. Benini, Fault tolerance overhead in network-on-chip flow control schemes, in *Proceedings of the 18th Annual Symposium on Integrated Circuits and System Design* (Florianolpolis, Brazil, 04–07 Sept. 2005), pp. 224–229
29. A.B. Ahmed, High-throughput architecture and routing algorithms towards the design of reliable mesh-based many-core network-on-chip systems, Ph.D. Thesis, Graduate School of Computer Science and Engineering, University of Aizu
30. Y. Tar, G.L. Frazier, High-performance multiqueue buffers for VLSI communication switches, in *15th Annual International Symposium on Computer Architecture* (1988), pp. 343–354
31. C.J. Glass, L.M. Ni, The turn model for adaptive routing, in *Proceedings of 19th Annual International Symposium Computer Architecture* (1992), pp. 278–287
32. A.B. Ahmed, A. Ben Abdallah, ONoC-SPL customized network-on-chip (NoC) Architecture and prototyping for data-intensive computation applications, in *IEEE Proceedings of The 4th International Conference on Awareness Science and Technology* (2012), pp. 257–262
33. *Altera design software*, http://www.altera.com/
34. Nios II processor, http://www.altera.com/literature/lit-nio2.jsp

Chapter 5
Advanced Multicore SoC Interconnects

Abstract Next-generation multicore SoC architectures are expected to combine hundreds of tiny cores integrated together to satisfy the power and performance requirements of large complex applications. As the number of cores continues to increase, the employment of low-power and high-throughput on-chip interconnect fabrics become imperative. This chapter describes the architecture and design of two emerging multicore SoC interconnects to overcome the limitations of the conventional (two-dimensional) multicore SoC on-chip interconnect. First, we present the architecture and design of three-dimensional interconnect, which promises a good opportunity for chip architects by porting the 2D-NoC to the third dimension. Second, we describe a mesh-based phototonic on-chip interconnect based on an energy-efficient non-blocking optical switch and contention-aware routing mechanisms.

5.1 Introduction

Emerging applications are getting more and more complex, demanding good architecture to ensure a sufficient bandwidth for any transaction between memories and cores as well as communication between different cores on the same chip. Because of these and other factors, 2D-NoC interconnect become not a suitable candidate for future large-scale many-core SoCs that are expected to accommodate hundreds of cores. More specifically, the limitation of the 2D-NoC paradigm comes from the high diameter that conventional 2D-NoC suffers from. The network's diameter is the number of hops that a flit traverses in the longest possible minimal path between a source–destination pair.

In 2D-NoC, if a given packet traverses a large number of hops to reach its destination, the communication latency will be long and consequently the throughput will be low. In other words, large network diameter has a negative impact on the worst-case routing latency in the system.

The seek for optimizing 2D-NoC-based architecture becomes more and more necessary, and many researches have been conducted to achieve this goal in various approaches, such as developing fast routers [1–5] or designing new high-throughput, and low latency network topologies [6–8]. One of these proposed solutions was

© Springer Nature Singapore Pte Ltd. 2017
A. Ben Abdallah, *Advanced Multicore Systems-On-Chip*,
DOI 10.1007/978-981-10-6092-2_5

porting the 2D-NoC architecture to the third dimension [9]. In the past few years, 3D-ICs have attracted a lot of attention as a potential solution to resolve the interconnect bottlenecks. A 3D chip is a stack of multiple device layers with direct vertical interconnects tunneling through them [10, 11].

So far, the achieved researches in this area have shown that 3D-ICs can achieve higher packing density due to the addition of a third dimension to the conventional two-dimensional layout; thanks to the reduced average interconnect length, 3D-ICs can achieve higher performance. Besides this important benefit, this reduction of total wiring, a lower interconnect power consumption can be obtained [12, 13], not to forget that circuitry is more immune to noise with 3D-ICs [9]. This may offer an opportunity to continue performance improvements using CMOS technology with smaller form factors, higher integration densities, and supporting the realization of mixed-technology chips [14]. As Topol [13] stated, 3D-IC can improve the performance even in the absence of scalability.

3D-NoC architecture responds to the scaling demands for future multicore and many-core SoCs, exploiting the short vertical links between the adjacent layers that can clearly enhance the system performance. This combination is expected to provide a new horizon of NoC and IC designs in general.

One of the important design steps that should be taken into consideration while designing a 3D-NoC is to implement an efficient router since it is the backbone of any NoC architecture. The router's performance depends on many factors and techniques such as the traffic pattern, the router pipeline design, and the network topology. As Feihui [15] stated, among these three factors we have less control over the traffic patterns compared with the topology and the pipeline design. Following this logic and assuming the topology choice was already taken, one of the most important router enhancements that can be done is to improve the pipeline design. By reducing the pipeline delay via pipelining optimization, not only we decrease the per-hop delay, but also the whole network latency will be reduced. On the other hand, the pipeline design is strongly associated with the adopted routing algorithm. Routing is the process of determining the path that a flit should take between one-source and one-destination nodes. Routing algorithm can be classified into minimal or non-minimal, depending on whether flits traveling from source to destination always use the minimal possible path or not.

Minimal routing schemes are shorter and require less complex hardware, but allowing non-minimal routes increases the path diversity and decreases the network congestion. Also the routing algorithms can be adaptive, where routing decisions are made based on the network congestion status and other information about network links or buffer occupancy of the neighboring nodes, or alternatively are deterministic.

There are a large number of sophisticated adaptive routing algorithms. However, they require more hardware and are difficult to implement. That is why deterministic routing schemes have been adopted for 3D-NoC designs. One of the well-used routing schemes used in 3D-NoCs is the Dimension-Order Routing (DOR) XYZ algorithm. XYZ is a simple scheme, easy to implement, and free of deadlock and lifelock. But on the other hand, it suffers from a non-efficient pipeline stage usage. This can introduce an additional packet latency which has an important effect on the router delay and eventually on the system overall performance. Enhancing this algorithm while keeping its simplicity may improve the system performance by reducing the packet delay.

A 2D-NoC, named OASIS-NoC, was presented in [16–18]. Although this architecture has its advantages over the shared-bus-based systems, it has also several limitations such as high power consumption, high-cost communication, and low throughput.

The presented 3D-OASIS-NoC (3D-ONoC) is based on a so-called Look-Ahead-XYZ (LA-XYZ) routing algorithm [19]. This algorithm improves the router pipeline design by parallelizing some stages while taking advantage at the same time of the simplicity of the conventional XYZ. As a result, this routing scheme aims to enhance the router performance thereby achieving a low-latency design.

5.2 Three-Dimensional On-Chip Interconnect

As we stated in Chap. 1, the number of transistors kept increasing along the past few decades, which made shrinking the chip size while maintaining high performance possible. This technology scaling has allowed Systems-on-Chip (SoCs) to grow continuously in component count and complexity, which significantly led to some very challenging problems, such as power dissipation and resource management [20, 21].

As moving to deep submicron technology poses real design and manufacturing problems, 3D integration becomes an attractive option to meet power and performance demands. By stacking dies or wafers we can reduce the wire length. As a result, the performance is increased and the power consumption is reduced. Thus, the on-chip interconnection network plays a more and more important role in determining the performance and also the power consumption of the entire chip [22].

Based on a simple and scalable architecture platform, NoC connects processing cores, memories, and other custom designs together using switching packets on a hop-by-hop basis. The ultimate goal is to provide a higher bandwidth and higher performance. Figure 5.1a, b show some well-known architectures which are, respectively, Point-to-Point (P2P) and shared-bus systems. As shown in Fig. 5.1c, NoC architectures are based upon connecting segments (or wires) and switching blocks to combine the benefits of the two previous architectures while reducing their disadvantages, such as the large numbers of long wires and the lack of scalability in shared-bus systems.

5.2.1 3D-NoC Versus 2D-NoC

3D-NoC is a widely studied research topic, and many related works have been conducted in the past. Few of them focused on the benefits of the 3D-NoC architecture over the traditional 2D-NoC design. Feero [23] showed that 3D-NoC has the ability

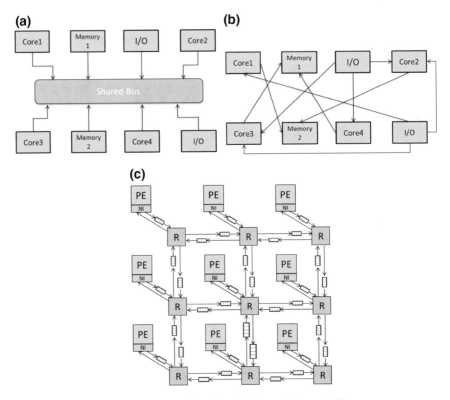

Fig. 5.1 SoC interconnection types: **a** Shared bus, **b** Point-to-Point, **c** NoC

to reduce latency and the energy per packet by decreasing the number of hopes by 40% which is a basic and important factor to evaluate the system performance [23]. Pavlidis [24] analyzed the zero-load latency and power consumption, and demonstrated that a decrease of 62 and 58% in power consumption can be achieved with 3D-NoC when compared to a traditional 2D-NoC topology for a network size of $N=$ 128 and $N=256$ nodes, respectively, where N is the number of cores connected to the network. This power consumption reduction can simply be related to the reduction of a number of hops, since a flit has less hops to traverse to go from one source to its destination, and that includes less buffer access, less switch arbitration, and less link and crossbar traversal. All of these factors will eventually lead to decrease in the power consumption.

Another part of previous works is focused on the router architecture. For example, Li [25] has modified the conventional 7 × 7 3D router using a shared bus as a communication interface between the different layers of the router, to create a *3D-NoC-Bus Hybrid* router. This kind of routers reduces in fact the number of ports in each router from 7 to 6, but on the other hand flits wishing to travel from one layer to another should compete the access to the shared bus, since it is the only interlayer communication interface. This may lead to undesirable performance degradation especially under a heavy interlayer traffic.

Yan [26], also proposed another architecture for the 3D-router, by implementing all the vertical links into a single 3D-crossbar. In this case, the router has only five ports since we do not need any more additional ports for the vertical connections. This technique reduces the interlayer distance and makes the travel between the different layers in one single hop possible. But this router also engenders a high router cost besides the implementation complexity of such router, which cannot be acceptable for some simple application that actually does not need such a complex router.

For all these facts, we adopted the conventional 7 × 7 3D-router, as it is the lowest cost among the other architectures and also the simplest to implement showing several properties like regularity, concurrent data transmission, and controlled electrical parameters [27, 28]. All the benefits are acquired while making sure that this low-cost and simple implementation does not affect the performance of our system.

5.2.2 Routing Algorithms

Many routing algorithms have been proposed for MCSoC systems but most of them focus only on 2D network topologies. Also, among all the studies conducted for 3D-NoC, few of them focused on routing algorithms. Among the few proposed ones, there are some custom routing schemes that aim to reduce the power consumption and thermal power which is a very challenge design for 3D-NoC systems. For instance, Ramanujam [29] presented an oblivious routing algorithm called randomized partially minimal (RPM) that aims to load balance the traffic along the network improving the worst-case scenario. RPM sends packets to a random layer first, and then route them along their X and Y dimensions using either XY or YX routing with equal probability. Finally, packets are sent to their final destination along the Z dimension.

In a quite similar technique, Chao [30] addressed the thermal power problem in 3D-NoC, which is one of the most important issues in the 3D-NoC designs. Starting from the fact the upper layer in the network detains the highest thermal power in the design, they proposed a thermal aware downward routing scheme that sends first the traffic to a downer layer, routes along the X and Y dimension before sending the packets back up to their destination layer. This technique avoids communication in upper layers, where the thermal power is more important than the downer ones, and then may reduce the overall thermal power in the design, thus ensuring thermal safety while guaranteeing less performance impact from temperature regulation.

Both of these two routing algorithms have their advantages in terms of load balancing and thermal power reduction. But the routing used is not minimal, which affect in a direct way the number of hops. By adopting a non-minimal routing, the packet delay may increase in the system, especially when we talk about a large number of connected nodes.

To ensure a minimal path for flits when traveling the network while making the routing as simple as possible, the majority of the remaining 3D-NoC systems have been using the conventional minimal Dimension-Order Routing (DOR) XYZ routing scheme. Other introduced a routing scheme based upon XYZ such as the case of *Tyagi* in [31] who extended a previous routing algorithm [32] called *BDOR* designated for 2D-NoC. *BDOR* forwards packets in one of two routes (XY- or YX-orders), depending on relative position of a source–destination pair, and that aims to improve the balance of paths along the network also when taking into account the destination.

XYZ routing scheme, and all the routing algorithms based upon it, is presented as a vertically balanced routing algorithm which has the best performance, since it is simple to implement, it is free of deadlock and lifelock, and also because packet ordering is not required [30, 33, 34]. On the other hand, it cannot always make the best use of each pipeline stage, for the simple reason that since the Switch Allocation stage (SA) is always dependent on the previous Routing Calculation (RC) one. This dependency can be explained by the fact that SA stage needs information about the desired output port calculated from the RC stage, where the incoming flits should go through in order to pass to the next neighboring node. To solve this problem in 2D-NoC systems using the Dimension-Order Routing (DOR) XY routing scheme, a smart pipeline design can be adopted with the help of some advanced techniques like look-ahead routing [31]. This kind of routing has been used to reduce the pipeline stages in the router, by parallelizing some of these stages, then reducing the router delay, and then enhancing the system performance. Look-ahead routing has indeed been used with 2D-NoC but it has not been adopted for 3D network-on-chip architectures before.

A second problem that can be seen with a lot of conventional routers using XYZ-based routing schemes is in case of no-load traffic and when the input buffer is empty, the flit entering the router should be first stored in the input buffer before advancing the next RC stage even there is no any flit under process in the next stages. This unnecessary stall will increase the packet latency in the router, and its associated power consumption, adding a performance overhead to the whole system even in a light traffic case where the system is supposed to have a close-to-optimal performance since there is no congestion that may increase the latency. In order to face this problem, a technique called no-load bypass is used [35]. This technique allows the flit to advance to the RC stage in case where the buffer is empty.

Previously in [36], a part of this research has been including architecture of a 3D network-on-chip architecture (named 3D-OASIS-NoC) based on a previously designed 2D-OASIS-NoC. The design's performance was evaluated using a simple application that randomly generates flits and sends them along the network. But real application could not be evaluated due to the absence of some components in the design such as the network interface. For that reason, a network interface has been added to 3D-ONoC, the optimized version of 3D-OASIS-NoC, in order to make our system able to be evaluated with our real selected target applications (JPEG encoder and Matrix Multiplication).

In this chapter we present a complete architecture and design of 3D-OASIS-NoC. Also evaluation results are presented using real applications (JPEG encoder and Matrix Multiplication). We provide more details about the different components of 3D-OASIS-NoC including a new Look-ahead-XYZ routing scheme (LA-XYZ) and its ability to take advantage of the simplicity of the conventional XYZ algorithm, while improving the pipeline design of the 3D-NoC router and enhancing the overall performance. The look-ahead routing scheme means that each flit additionally carries one hot encoded *next-port* identifier used by the downstream router. The no-load bypass technique is also associated with LA-XYZ in order to get more pipeline improvement.

5.2.3 Topology Design

3D-ONoC is a scalable network-on-chip based on *Mesh* topology. The packets are forwarded among the network using *Wormhole-like* switching policy and then routed according to *Look-Ahead-XYZ* routing algorithm (LA-XYZ). Many topologies exist for the implementation of NoCs; some are regular (*Torus, tree-based*) and other irregular topologies are customized for some special application. We choose the Mesh topology for this design, thanks to its several properties like regularity, concurrent data transmission, and controlled electrical parameters [27, 28].

Figure 5.2 shows a configuration example of $4 \times 4 \times 4$ 3D-ONoC design. We can see in this figure that different layers are linked between each other via interlayer channels. On the other side, each layer is composed of different switches which are connected to each other using some intra-layer links; each one of them is connected to one single processing element.

Code 5.1 illustrates the RTL (in Verilog HDL) code of the 3D-ONoC top module that defines the mesh topology. The z-loop, y-loop, and x-loop are used to define the dimensions of 3D-NoC, while the internal i-loop (line 17) is used to define the different input and output ports for each direction. For example, $i = 0$ refers to the local port, where the outputs and inputs of this port will be allocated later to the attached PE.

Taking the example of the *down port* (line 31–39), the output and input of this port are allocated to the UP port of the router situated just below the current router, which means the one in the downer layer. As it will be explained later, the unused ports should be eliminated in order to reduce the area and power consumption. Continuing with the same *down port*, it should be disabled when the router is located at the bottom of the topology, which means when $z - pos = 0$. In this case, as it is illustrated in Code 5.1 (line 32–35), net-data-in and net-stop-in are assigned to 0.

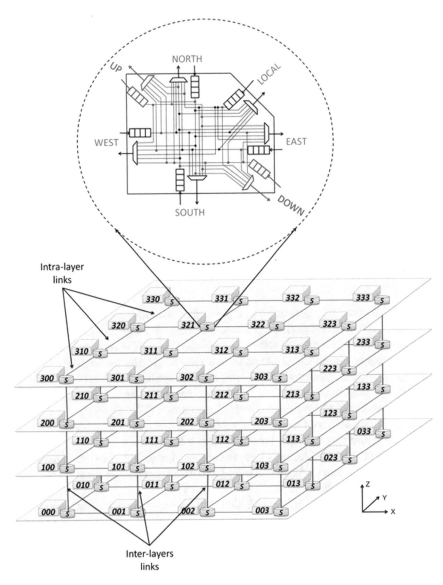

Fig. 5.2 Configuration example of a 4 × 4 × 4 3D-ONoC based on mesh topology

Listing 5.1 Verilog HDL code defining the topology

```
1  generate
2  //z loop
3  for (z_pos=0; z_pos<Z_WIDTH; z_pos=z_pos+1) begin:z_loop
4
5      //y loop
6      for (y_pos=0; y_pos<Y_WIDTH; y_pos=y_pos+1) begin:y_loop
7
8          //x loop
```

```
9    for (x_pos=0; x_pos<X_WIDTH; x_pos=x_pos+1) begin:x_loop
10
11        router #(NOUT, FIFO_DEPTH, FIFO_LOG2D, FIFO_FULL_LVL) rtr
             (.clk(clk), .reset(reset),
12             .data_in(net_data_in[x_pos][y_pos][z_pos]), .
                  data_out(net_data_out[x_pos][y_pos][z_pos]),
13             .stop_in(net_stop_in[x_pos][y_pos][z_pos]), .
                  stop_out(net_stop_out[x_pos][y_pos][z_pos]),
14             .xaddr(x_pos['L2NET_SIZE-1:0]), .yaddr(y_pos['
                  L2NET_SIZE-1:0]), .zaddr(z_pos['L2NET_SIZE
                  -1:0]));
15
16        //set up inter-router connections with correct boundary
             conditions
17        for (i=0; i<NOUT; i=i+1) begin:i0
18
19            //tile interface of router
20            if(i==0) begin
21            assign net_data_in[x_pos][y_pos][z_pos]['WIDTH*(i+1)-1:'
                  WIDTH*i] = data_in[('WIDTH*X_WIDTH*z_pos*Y_WIDTH)+('
                  WIDTH*X_WIDTH*y_pos)+ ('WIDTH*(x_pos+1))-1: ('WIDTH*
                  X_WIDTH*z_pos*Y_WIDTH)+('WIDTH*X_WIDTH*y_pos)+('
                  WIDTH*x_pos)];
22            assign data_out[('WIDTH* X_WIDTH*z_pos*Y_WIDTH)+('WIDTH*
                  X_WIDTH*y_pos)+('WIDTH*(x_pos+1))-1: ('WIDTH*X_WIDTH
                  *z_pos*Y_WIDTH) +('WIDTH*X_WIDTH*y_pos)+ ('WIDTH*
                  x_pos)] = net_data_out[x_pos][y_pos][z_pos]['WIDTH*(
                  i+1)-1:'WIDTH*i];
23
24            assign net_stop_in[x_pos][y_pos][z_pos][i] = stop_in[(
                  X_WIDTH* z_pos * Y_WIDTH)+(X_WIDTH*y_pos)+x_pos];
25            assign stop_out[(X_WIDTH* z_pos * Y_WIDTH)+(X_WIDTH*
                  y_pos)+x_pos] = net_stop_out[x_pos][y_pos][z_pos][i
                  ];
26            end
27    ...
```

5.2.4 Switching Policy

Considered as a very important choice for any NoC design, switching establishes the type of connection between any upstream and downstream node. It is important to deploy an efficient switching policy to ensure less blocking communication while trying to minimize the system complexity. When it is related to packet switching, three main switching policies have been mostly used for NoC: *Store and Forward (SAF)*, *Virtual Cut Through (VCT)*, and *Wormhole (WH)* [37].

Listing 5.2 Verilog-HDL code defining the flit structure

```
1  // Flit structure
2  `define DATA        37:0
3  `define TAIL        0
4  `define NEXT_PORT   7:1
5  `define XDEST       10:8
6  `define YDEST       13:11
7  `define ZDEST       16:14
8  `define DATA        37:17
```

3D-ONoC adopts *wormhole-like* switching and virtual-cut-through forwarding method. The forwarding method which is chosen in a given instance depends on the level of packet fragmentation. For instance, each router in 3D-ONoC has input buffers which can store up to four flits by default. When a packet is divided into more than four flits, 3D-ONoC chooses virtual-cut-through switching. When packets are divided into less than four flits, the system chooses wormhole. In other words, when the buffer size is greater than or equal to the number of flits, virtual-cut-through is used, but when buffer size is less than or equal to the number of flits, wormhole switching is employed. By combining the benefits of both switching techniques, packet forwarding can be executed in an efficient way while guaranteeing a small buffer size. As a result the system performance is enhanced while maintaining a reasonable area utilization and power consumption.

5.2.4.1 Flit Format Design

Figure 5.3 shows the 3D-ONoC flit format. The first bit indicates the *tail* bit informing the end of the packet. The next seven bits are dedicated for the *next-port* that will be used by the *Look-Ahead-XYZ* routing algorithm to define the direction of the next downstream neighboring node where the flit will be sent to. Then, three bits are used to store destination information of each: *xdest*, *ydest*, and *zdest*. Having three bits for each destination field allows the network to have a maximum size of $8 \times 8 \times 8$ 3D-ONoC. But if the network size needs to be extended, the addresses fields may also be increased to accommodate a larger network size. Finally, the remaining 64 bits are dedicated to store the payload. Since 3D-ONoC is targeted for various applications, the payload size can be easily modified in order to respect the requirements of some specific applications. Code 5.2 shows the structure of the 3D-ONoC flit. In addition, as we previously stated, the architecture does not provide for a separate head flit and every flit, therefore, identifies its destination X, Y, and Z addresses and carries an additional single bit to indicate whether it is a tail flit or not.

5.2.5 3D-NoC Router Architecture Design

The router is considered as the backbone element in the whole 3D-ONoC design. The 3D-ONoC router architecture is based upon the 5×5 2D-ONoC router where, as shown in Fig. 5.2, each switch has a maximum number of 7-input by 7-output port, where four ports are dedicated to connect to the neighboring routers in north, east,

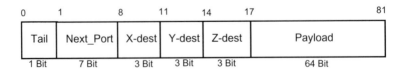

Fig. 5.3 3D-ONOC flit format

south, and west directions using the intra-layer links. One port is used to connect the router to the local computation tile where the packet can be injected into or ejected from the network. The remaining two ports are added to connect the switch to the upper and downer layers to ensure the interlayer communication.

As we previously stated, the number of ports depends on the position of the switch in the design, since we have to eliminate any unused links that have no connections with other switches in order to reduce power consumption. For example, as it is depicted in Fig. 5.2, switch-000 have only four connected ports (north, east, up, and local) and the remaining three ports (south, west, and down) have been disabled since there are no connections to any neighboring routers along those directions. Figure 5.4 represents 3D-ONoC router architecture and that the routing process at each router can be defined by three main pipeline stages: buffer writing (BW), routing calculation and switch allocation (RC/SA), and the crossbar traversal stages (CT). Observing the Verilog HDL code for the *Router* module depicted in Code 5.3, 3D-ONoC contains seven *input-port* modules for each direction represented in *input-port* module in line 4. This seven-module allocation is defined by the i-loop in line 2, where each value of i refers to the seven directions (local, north, east, south, west, up, down), and *NOUT* parameter in line 2 refers to the number of ports. The outputted *sw-req* signal defining the input port asking the grant and the output port requested defined by the *port-req* signal is sent from the seven-input port to be an input port for the switch allocator as shown at lines 19 and 20 of Code 5.3.

In addition to the *switch allocator*, the *crossbar* module is also defined (lines 22–25). The crossbar circuit takes as input the *sw-cntrl* from the switch allocator and *data-in* coming from the seven input ports.

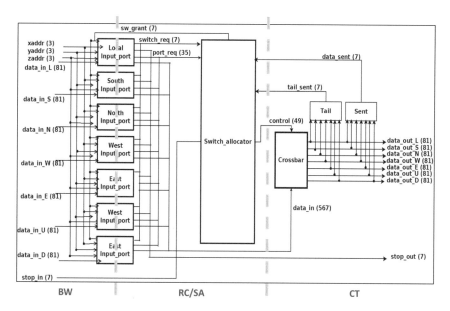

Fig. 5.4 3D-ONoC pipeline stages: buffer writing (BW), routing calculation and switch allocation (RC/SA) and crossbar traversal stage (CT)

Listing 5.3 Verilog-HDL Code for Router

```
1  //instantiate input ports
2  for (i=0; i<NOUT; i=i+1) begin:il
3
4      input_port #(NOUT, FIFO_DEPTH, FIFO_LOG2D, FIFO_FULL_LVL) ip
5              (.clk(clk), .reset(reset),
6          .data_in(data_in['WIDTH*(i+1)-1:'WIDTH*i]),
7          .data_out(cbar_data_in['WIDTH*(i+1)-1:'WIDTH*i]),
8          .sw_req(sw_req[i]), .port_req(port_req[NOUT*(i+1)-1:
               NOUT*i]),
9          .sw_grant(sw_grant[i]), .stop_out(stop_out[i]),
10         .xaddr(xaddr), .yaddr(yaddr), .zaddr(zaddr));
11
12      assign data_sent[i] = |data_out['WIDTH*i+'NEXT_PORT_END:'
            WIDTH*i+'NEXT_PORT_START];
13      assign tail_sent[i] = data_out['WIDTH*i];
14
15     end
16  endgenerate
17
18    sw_alloc #(NOUT) sw_allc(.clk(clk), .reset(reset),
19    .sw_req(sw_req), .stop_in(stop_in), .data_sent(data_sent), .
            tail_sent(tail_sent),
20    .port_req(port_req), .grant_out(sw_grant), .sw_cntrl(sw_cntrl));
21
22    crossbar #(NOUT, NOUT, 'WIDTH) cbar(.clk(clk), .reset(reset),
23                    .cntrl(sw_cntrl),
24                    .data_in(cbar_data_in),
25                    .data_out(data_out));
```

Now we analyze each component of the switch separately. Starting with the *input port*, the *switch allocator*, and finally *crossbar* module.

5.2.5.1 Input-Port Module Design

Starting with the *input-port* module represented in Fig. 5.5 (and where the Verilog code is represented in Code 5.4), each one of the seven modules is composed of two main elements: *Input buffer* and the *Route* module.

Listing 5.4 Verilog-HDL Code for Input port

```
1  //instantiate FIFO
2    fifo #(NOUT, FIFO_DEPTH, FIFO_LOG2D, FIFO_FULL_LVL) ff
3            (.data_in(data_in), .data_out(fifo_data_out),
4          .second_item_nextport(second_fifo_nextport),
5        .enqueue(enqueue), .dequeue(sw_grant),
6        .stop_out(stop_out), .nearly_empty(fifo_nearly_empty),
7          .empty(fifo_empty),
8        .clk(clk), .reset(reset));
9
10  //instantiate look-ahead routing module
11    route #(NOUT) rr
12            (.xdest(fifo_data_out['XDEST]), .ydest(fifo_data_out['
               YDEST]),.zdest(fifo_data_out['ZDEST]),
13          .xaddr(xaddr), .yaddr(yaddr), .zaddr(zaddr),
14          .nextport(fifo_data_out['NEXT_PORT]), .new_nextport(
               lookahead_route));
```

Incoming 81-bit flits *data-in* from different neighboring switches, or from the connected computation tile, are first stored in the *Input buffer* and waiting to be processed.

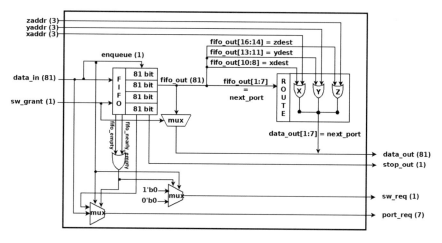

Fig. 5.5 Input-port module architecture

This step is considered as the first pipeline stage of the flit's life cycle (BW). As it is illustrated in Code 5.5, arbitration between different flits is managed using FIFO queue technique. Each input buffer has by default four as depth, which means that it can host up to four 81-bit flits. Buffers occupy a significant portion of router area but can imply also increase in overall performance.

Listing 5.5 Verilog-HDL Code for Input-FIFO buffer

```
1    always @(posedge clk) begin
2        if (!reset) begin        //If out of reset
3            if (enqueue) begin //Write a flit to the buffer
4                fifo[tail_ptr] <= data_in;
5                tail_ptr <= tail_ptr + 1;
6            end
7            if (dequeue) begin //Read a flit from the buffer
8                head_ptr <= head_ptr + 1;
9            end
10   //nearly full signal = stop_out,
11           if (((tail_ptr + FULL_LVL[LOG2D-1:0] + 1'b1)==head_ptr
                 ) && enqueue && !dequeue)begin
12               stop_out <= 1'b1;
13           end
14           if (((tail_ptr + FULL_LVL[LOG2D-1:0])==(head_ptr+1'b1)
                 ) && !enqueue && dequeue)begin
15               stop_out <= 1'b1;
16           end
17           if ((tail_ptr + FULL_LVL[LOG2D-1:0])==head_ptr)begin
18               if ((enqueue && !dequeue) || (!enqueue && dequeue)
                     )begin
19                   stop_out <= 1'b0;
20               end
21   ...
```

After being stored, the flit is fetched from the *FIFO* buffer and advanced to the next pipeline stage (RC/SA). The destination addresses (*xdest*, *ydest*, and *zdest*) are then decoded in order to extract the information about the destination address in addition to the *next-port* pre-calculated in the previous upstream node. Those values are then sent to the *Route* circuit where La-XYZ routing scheme is executed to determine the *new next-port* direction for the next downstream node. At the same time, the *next-port* identifier is also used to generate the request for the *switch allocator* asking for grant to use the selected output port via *sw-req* and *port-req* signals.

As we stated in the previous section, 3D-ONoC uses look-ahead routing scheme *LA-XYZ* for fast routing. This scheme is based upon the dimension-order (DOR) X–Y–Z static routing algorithm, where the X,Y, and Z coordinates are satisfied in order. X–Y–Z routing is presented as the vertically balanced routing algorithm which has the best performance, since it is simple to implement, it is free of deadlock and livelock, and also because packet ordering is not required. In addition to that each flit additionally carries one hot encoded *next-port* identifier used by the downstream router. Since *LA-XYZ* is based upon *XYZ* routing, it is considered also as a minimal routing where each flit from any source and destination pair traverses the minimal number of hops.

5.2.5.2 Semi-adaptive Look-Ahead Routing

To understand better how the *next-port* is decided, we designed the Verilog HDL code depicted in Code 5.6. As it is shown in this code (lines 1–12), the routing decision starts first by finding the next node's address. It is done by evaluating the actual *next-port* fetched from the flit, which gives a hint about which neighboring node the flit is going to be routed to and eventually knowing its exact address by incrementing *xaddr* or *yaddr* or *zaddr*. Depending on the resulted next address from the later step, the new *next-port* can be determined. As demonstrated between lines 15 and 31 in Code 5.6, *LA-XYZ* compares the resulted next node's address (*next-xaddr*, *next-yaddr*, and *next-zaddr*) and the destination addresses (*xdest*, *ydest*, and *zdest*). At the end of the execution of this comparison, the new *next-port* (defined by *route* in Code 5.6) can be determined and then embedded in the flit back again to be sent to the next node as Fig. 5.5 illustrates.

Listing 5.6 Verilog HDL implementation of LA-XYZ routing algorithm.

```
1      //assign next addresses
2          if (nextport == 'EAST) next_xaddr = xaddr + 1'b1;
3        else if (nextport == 'WEST) next_xaddr = xaddr - 1'b1;
4        else next_xaddr = xaddr;
5
6          if (nextport == 'NORTH) next_yaddr = yaddr + 1'b1;
7        else if (nextport == 'SOUTH) next_yaddr = yaddr - 1'b1;
8        else next_yaddr = yaddr;
9
10          if (nextport == 'UP) next_zaddr = zaddr + 1'b1;
11        else if (nextport == 'DOWN) next_zaddr = zaddr - 1'b1;
12        else next_zaddr = zaddr;
13
14      //evaluate next port
15      if (next_xaddr == xdest)
16      begin  if (next_yaddr == ydest)
17        begin        if (next_zaddr == zdest) route = 'SELF;
18            else begin if(next_zaddr < zdest) route = 'UP;
19                    else route = 'DOWN;
20                end
```

```
21                    end
22          else    begin
23                          if(next_yaddr  <  ydest)  route  =  `NORTH;
24                          else  route  =  `SOUTH;
25                    end
26          end
27          else    begin
28              if  (next_xaddr  <  xdest)  route  =  `EAST;
29                  else  route  =  `WEST;
30          end
31          end
```

If we take a look at Fig. 5.2, and assume for example that a flit coming from switch-200 enters switch-201 (where the *xaddr*, *yaddr*, and *zaddr* addresses are defined by 001, 000, and 001, respectively) trying to reach its destination node switch-313 (where the *xdest*, *ydest*, and *zdest* addresses are defined by 011, 001, and 011, respectively). This flit caries "EAST" as a *next-port* identifier pre-calculated in the previous node (switch-200). According to the first phase of the LA-XYZ algorithm, *next-xaddr* = *xaddr* + 1 which is the x-address of switch-202. In the second phase of the algorithm, *next-xaddr* is then compared with *xdest*. The comparison result will determine "EAST" as *route* (the new *next-port* for switch-202) which will be re-updated in the flit.

In order to enable the bypass technique, two signals are issued from the buffer to give information about the buffer occupancy status. These two signals are *fifo-empty* and *fifo-nearly-empty*. When the *fifo-empty* signal is issued, it means that the input buffer is empty and when an incoming flit arrives at the input port, it does not need to be stored in the buffer. Then, overlap the buffering stage and advancing to the next stage (RC and SA).

5.2.5.3 Switch Allocator Design

The *sw-req* and *port-req* signals issued from each *input-port* module, and giving information about the desired output port, are transmitted to the *switch allocator* module to perform the arbitration between the different requests. When more than two input flits from different input ports are requesting the same output port at the same time, the *switch allocator* manages to decide which output port should be granted to which input port, and when this grant should be allocated. This process is done in parallel with the routing computation done in *Input port* to form the second pipeline stage.

As indicated in Fig. 5.6, the switch allocator circuit has two output signals: one is *sw-cntrl* and the second one is *grant-out*. sw-cntrl contains all the information needed by the crossbar circuit about the scheduling result as it is explained later. On the other hand, the *grant-out* is sent back to the *input-port* module and gives the grant to the appropriate input port to send its data to the crossbar before reaching its next neighboring node. Figure 5.6 shows that the switch allocator module is composed of two main components: *Stall-Go flow control* and *Matrix-Arbiter Scheduling*.

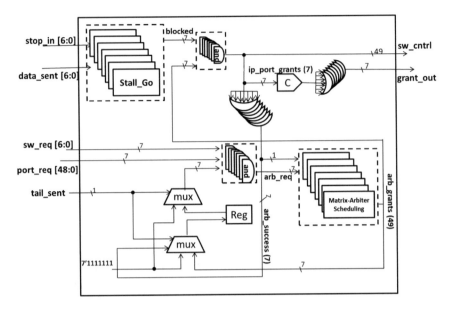

Fig. 5.6 Switch allocator architecture

5.2.5.4 Stall-Go Flow Control Architecture

Like the other flow control schemes, *Stall-Go* module manages the case of the buffer overflow. When the buffer exceeds its limitation on hosting flits (if the number of flits waiting for process is greater than the depth of the buffer), a flow control has to be considered to prevent from buffer overflow and eventually from packet dropping, thus allocating available resources to packets as they progress along their route. We chose *Stall-Go* flow control since it proves to be a low-overhead efficient design choice showing remarkable performance comparing with the other flow control schemes such as *ACK-NACK* or *Credit-based* flow control. Like the other flow control schemes, *Stall-Go* module manages the case of the buffer overflow. When the buffer exceeds its limitation on hosting flits (if the number of flits waiting for process is greater than the depth of the buffer), a flow control has to be considered to prevent from buffer overflow and eventually from packet dropping, thus allocating available resources to packets as they progress along their route. We chose *Stall-Go* flow control since it proves to be a low-overhead efficient design choice showing remarkable performance comparing to the other flow control schemes such as *ACK-NACK* or *Credit-based* flow control [38].

Listing 5.7 Verilog HDL of the state machine decision

```
1    always @(posedge clk) begin
2
3        if (!reset) begin
4            if ((state=='GO) && stop_in && data_sent)
5                state <= 'SENT1;
6            if (state=='SENT1) begin
```

```
7         if (stop_in && !data_sent)
8             state <= 'GO;
9         if (!stop_in && data_sent)
10            state <= 'STOP;
11    end
12
13 if ((state=='STOP) && stop_in) // stop_in = nearly_full
14    state <= 'GO;
15 end else
16    state <= 'GO;
17 end
18
19 assign blocked = ( ((state=='STOP) && !stop_in) || ((state=='
       SENT1) && !stop_in && data_sent) );
```

Stall-Go module, where the mechanism is represented in Fig. 5.7, uses two control signals: *nearly-full* and *data-sent*. *nearly-full* signal is sent to the upstream node indicating that the input buffer is almost full and only one slot is still available to host one last flit. After receiving this signal, the *FIFO* buffers suspend sending flits. The *data-sent* signal is issued when the flit is transmitted. Figure 5.8 represents the *Stall-Go* flow control state machine which aims to generate the *nearly-full* and *data-sent* signals. State *GO* indicates that the buffer is still able to host two or more flits. State *SENT* indicates that the buffer can host only one more flit, and finally when we

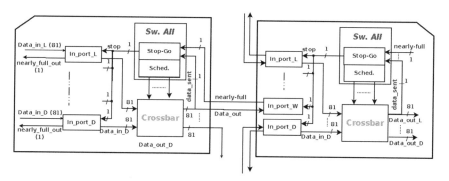

Fig. 5.7 Stall-Go flow control mechanism

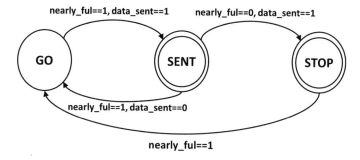

Fig. 5.8 Stall-Go flow control finite state machine

move to state *STOP*, it means that the buffer cannot store anymore flits. The state machine is generated as indicated in Code 5.7 that shows the Verilog-HDL code explaining the main state transitions using *nearly-full* and *data-sent* signals.

5.2.5.5 Matrix-Arbiter Scheduling Architecture

The second component is the scheduling module. As shown in Fig. 5.9, the input signals *sw-req* and *port-req* indicate the input ports demanding the access and which output ports are they requesting, respectively. Depending on these requests, the arbiter allocates the convenient output port to its demander. Since 3D-ONoC transmits only one flit in every clock cycle, then when two input ports or more are competing for the same output port, the presence of a scheduling scheme is required in order to prevent any possible conflict. The switch allocator in our design employs a least recently served priority scheme via the packet transmit layer. Thus, it can treat each communication as a partially fixed transmission latency [39, 40]. Matrix arbiter is used for a least recently served priority scheme.

In order to adopt matrix arbiter scheduling for 3D-ONoC, we implemented a 6×6 scheduling matrix. The scheduling module accepts all the requests from the different connected input ports and their requested output ports. Then it assigns priority for each request. In order to give the grant to the convenient input port, the scheduling module verifies the scheduling matrix, compares the priorities of the input ports competing for the same output port, and gives the grant to the one possessing the highest priority in the matrix. Following this basis, the scheduling module should make the input port, which got the last grant to use the completed output port, the lowest priority for the next round of arbitration, and then increases the priority of the rest of the remaining ports. When there are no requests, the priority is unchanged. Based on these assumptions, we are sure that every input port will be served and get the grant to use the output port in a fair way. Figure 5.9 illustrates a simple example of how our scheduling mechanism works. Each row of the matrix represents the competing input requests and their priorities. The scheduling module starts by examining the priorities of each input port request. After the highest priority input is served, the arbiter updates the scheduling matrix by making the request which got the last grant, the lowest priority for the next round of arbitration, by inverting its row and column.

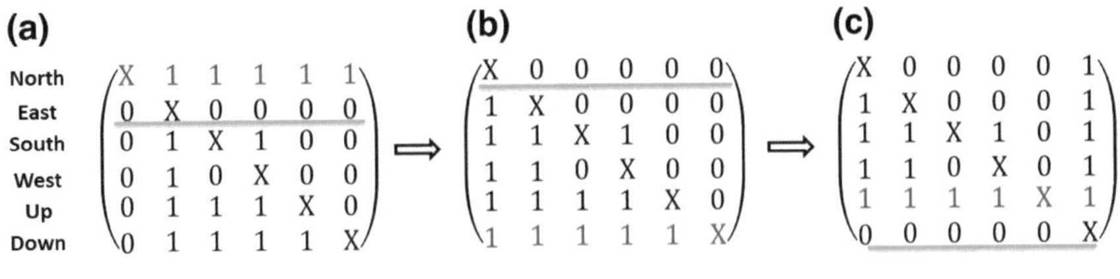

Fig. 5.9 Scheduling matrix priority assignment

The matrix shown in Fig. 5.9a illustrates the initial scheduling matrix where *north*, *up*, and *down* input ports are asking the grant to eject their flits to the *Local* port. Observing this figure, the *north* request (highlighted in red) has higher priorities compared with the remaining two requests. As a result the arbiter gives the grant to the *north* request. Then *north* becomes the lowest priority (as it is underlined by a green line) and the remaining two requests priorities are incremented. In the next round (Fig. 5.9b), *Down* seems to have a higher priority than the *Up* request. The arbiter then gives the grant to *Down* and makes its priority the lowest. Finally, as it is shown in Fig. 5.9c, the *Up* request having the highest priority among the others is giving the grant to eject its data to the requested output port. Code 5.8 depicts the Verilog HDL code for the implementation for the matrix arbiter.

Listing 5.8 Matrix Arbiter code

```
generate
for (i=0; i<SIZE; i=i+1) begin:o11
    for (j=0; j<SIZE; j=j+1) begin:i11
        if (j==i)
            assign pri[i][j]=request[i];
        else
        if (j>i)
            assign pri[i][j]=!(request[j]&&state[j*SIZE+i]);
        else
        assign pri[i][j]=!(request[j]&&!state[i*SIZE+j]);
        end
    assign grant[i]=&pri[i];
end
endgenerate

generate
for (i=0; i<SIZE; i=i+1) begin:o12
    for (j=0; j<SIZE; j=j+1) begin:i12
        assign new_state[j*SIZE+i]=(success&&((state[j*SIZE+i
            ]&&!grant[j])||(grant[i])))||(!success&&state[j*
            SIZE+i]);
    end
end
endgenerate

always@(posedge clk) begin
    if (reset) state<=-1;
    else begin
    if (|request) state<=new_state;
    end
end
```

Listing 5.9 Code for Crossbar circuit

```
//crossbar.v
generate
    for (i=0;i<NOUT;i=i+1) begin:output_loop
    mux_out #(NIN, WIDTH) cbar_mux(.cntrl(cntrl_reg[NIN*(i+1)
        -1:NIN*i]), .data_in(data_in), .data_out(data_out[
        WIDTH*(i+1)-1:WIDTH*i])));
    end
endgenerate

//mux_out
generate
    //loop over each bit of data
    for (i=0;i<WIDTH;i=i+1) begin:bit_loop
        assign data_out[i] = mux(cntrl, data_bits[i]);
```

```
13          //loop over each input channel
14          for (j=0;j<n_in;j=j+1) begin:input_loop
15              assign data_bits[i][j] = data_in[WIDTH*j+i];
16          end
17      end
18  endgenerate
19
20  function mux;
21      input [n_in-1:0] cntrl;
22      input [n_in-1:0] data_in;
23      integer i;
24
25  begin
26      mux = 0;
27      for (i=0; i<n_in; i=i+1) begin
28          if(cntrl[i] == 1'b1) mux = data_in[i];
29      end
30  end
31  endfunction // mux
```

5.2.5.6 Crossbar Design

The switch allocator sends the issued *control* signal to the crossbar circuit to complete the third and final Crossbar Traversal pipeline stage (CT), where information about the selected input port and the *next-port* are embedded, and then stored in the *sw-cntrl-reg* register as it is shown in Fig. 5.10. After that, the crossbar fetches these information, receives the data from the FIFO buffer of the selected input port, and

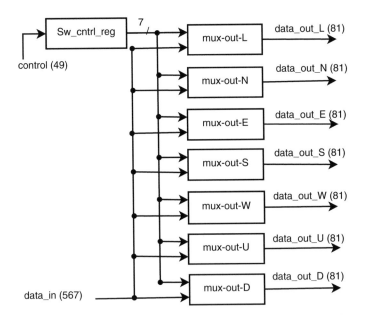

Fig. 5.10 Crossbar circuit

then allocates the appropriate channel for transmission to the decoded *next-port*. Finally, the crossbar sends the flit to its destination as illustrated in Fig. 5.10. When all the flits are transmitted, the *tail* bit informs the switch allocator via a *tail-sent* signal that the packet transmission is completed and can free the used channel so it can be exploited by another packet. Code 5.9 depicts the Verilog HDL code for the implementation for the crossbar circuit.

5.2.6 Network Interface Architecture

In order to enable real applications to be run on 3D-ONoC system, a Network Interface (NI) was added to every router as a medium interface between the different PEs (cores, memory, I/O, etc.). JPEG encoder application [41] was used for evaluating the system performance. For this, both *Transmitter* and *Receiver*-NI in every switch are designed. The packet size is set to 99-bit (3-bit flits). Each flit contains 17 bits defining the routing information (*xdst*, *ydst*, *zdst*, *next-port*, and *tail*) and the remaining 16 bits are dedicated for the payload.

Listing 5.10 Verilog-HDL sample code for the sending NI

```
module NI_02_send (clk, rst, enable, data_in, flit);
   input                clk, rst;
   input                enable;
   input [23:0]         data_in;
   output reg [32:0]        flit;
    always @(state)begin
      case(state)
    'f0:begin
        if(cntrl) begin
       next_state <= 'f1;
         flit <= 33'hz;
        end
        else next_state <= 'f1;
   end
    'f1:begin
       next_state <= 'f3;
       flit 0          <= 'header;
       flit[7:1]       <= 'EAST;
       flit[16:8]   <= 'dest_03;
       flit[32:17] <= data_in[23:8];
   end
    'f2:begin
        if(cntrl) begin
       next_state <= 'f3;
       flit 0          <= 'header;
       flit[7:1]       <= 'EAST;
       flit[16:8]   <= 'dest_03;
       flit[32:17] <= data_in[23:8];
       end
       else next_state <= 'f2;
   end
    'f3:begin
       next_state <= 'f4;
       flit 0          <= 'header;
       flit[7:1]       <= 'EAST;
       flit[11:8]   <= 'dest_03;
       flit[24:17] <= data_in[7:0];
       flit[32:25] <= 0;
```

```
39    end
40    'f4:begin
41       next_state  <=  'f2;
42       flit 0        <=  'tail;
43       flit[7:1]     <=  'EAST;
44       flit[16:8]   <=  'dest_03;
45       flit[17]      <=  enable;
46       flit[32:18]  <=   0;
47    end
48    default:next_state  <=  'f0;
49       endcase
50    end
```

Figure 5.11 shows the architecture of the *Transmitter-NI* , and Fig. 5.12 shows the
architecture of the *receiver-NI*. The NI receives a 32-bit data from the JPEG module
that will be divided into two portions representing the payload of the two first flits
of the packet. The payload of the third flit contains the 10-bit control signal from the
JPEG module, and the remaining six bits are unused. As shown in Fig. 5.11, a *Control
Module* manages the fits generation. It adds the convenient destination addresses and
next-port direction to each flit, and marks the end of the packet by adding the (*tail*
bit to the third final flit. The generated flits are then injected into the network. The
Verilog HDL implementation of the *Transmitter-NI* is depicted in Code 5.10.

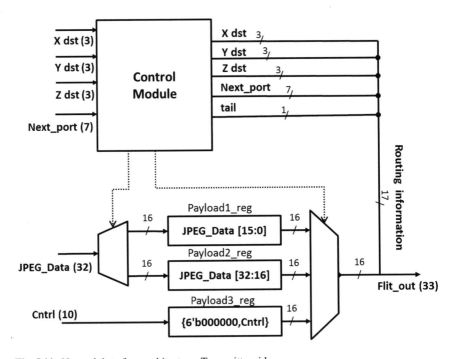

Fig. 5.11 Network interface architecture: Transmitter side

Listing 5.11 Verilog-HDL code for the receiving NI

```verilog
module NI_03_rec (clk, rst, flit, data_out, enable);

//input output
    input                   clk, rst;
    input [32:0]            flit;
    output reg [23:0]           data_out;
    output reg              enable;
    //reg
    reg [15:0]          data_high;
    reg [7:0]           data_low;
    reg                 ena;

    reg [1:0]           state;
    reg [1:0]           next_state;
    reg [32:0]      preflit;

    reg [23:0]          pre_data_out;
    //state
    always @(posedge clk)begin
        if(rst==1)state <= 'f0;
        else begin
        preflit <= flit;
        state <= next_state;
        end
    end

    //state
    always @(state or flit)begin
        case(state)

    'f0:begin
        if(flit!=preflit)begin
            next_state <= 'f1;
            data_high <= 0;
            data_low <= 0;
        end
        else next_state <= 'f1;
    end
    'f1:begin
        if(flit!=preflit)begin
            next_state <= 'f2;
            data_high <= flit[32:17];
        end
        else next_state <= 'f1;
    end
    'f2:begin
        if(flit!=preflit)begin
            next_state <= 'f3;
            data_low <= flit[24:17];
        end
        else next_state <= 'f2;
    end
    'f3:begin
        if(flit!=preflit)begin
            next_state <= 'f1;
            ena <= flit[17];
                pre_data_out <= {data_high, data_low};
        end
        else next_state <= 'f3;
    end
    default:next_state <= 'f0;
        endcase
    end
```

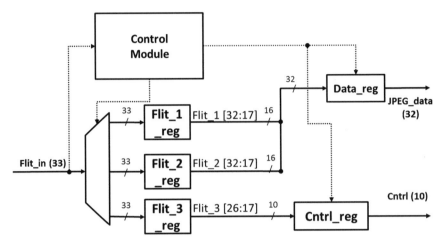

Fig. 5.12 Network interface architecture: Receiver side

On the other side, the *Receiver-NI* receives the incoming of three flits of each packet ejected from the network, and then stores them into three temporary registers. After that the 16-bit payloads of the first and second flit are fetched from the temporary registers, reassembled together and finally stored in the *Data-reg* register. Controlled by another *Control Module*, the complete 32 bits resulted in data and the 10 bits control signals are fetched and sent to their attached JPEG module after the complete packet is received.

The Verilog HDL implementation of the *Transmitter-NI* is depicted in Code 5.11. Based on this network interface, another one has been designed to satisfy the requirements of another application that we used for evaluating 3D-ONoC, which is matrix multiplication. We chose the matrix multiplication as one of our evaluating targets, since it is wildly used in scientific application. Due to its large multidimensional data array, it is extremely demanding in computation power and meanwhile, it is potential to achieve its best performance in a parallel architecture and does not involve synchronization [42]. All of these reasons make the matrix multiplication a very suitable application to evaluate 3D-ONoC and show its outperforming performance against 2D-ONoC.

5.2.7   3D-NoC Design Evaluation

In this section, we evaluate the hardware complexity of 3D-ONoC in terms of area utilization, power consumption (static and dynamic), and clock frequency. JPEG encoder [41] and matrix multiplication [42] applications were used. Execution time, the number of hops, and also the number of stall after the execution of the both of the applications are also analyzed. Comparison research is also performed with 2D-NoC architecture.

Fig. 5.13 Task graph of the
JPEG encoder

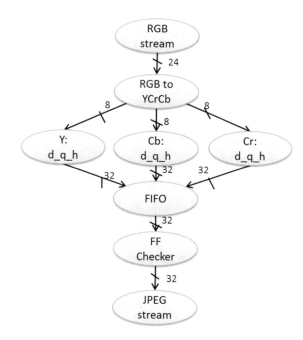

5.2.7.1   JPEG Encoder on 3D-ONoC

JPEG encoder application is a well-known application and is widely used for evaluating systems which expose a lot of parallelism. For instance, we took into consideration the tasks implementation shown in Fig. 5.13. For additional analysis, we
made further divisions to the *Y:d-q-h*, *Cb:d-q-h*, *Cr:d-q-h*, and *FIFO* modules, and
the resulted task graph is illustrated in Fig. 5.14. This extension aims to increase
the network size and deploy more parallel execution of the different modules of the
application, and then can take advantage of the scalability and the reduced number of
hops in the design. As we analyze the modified task graph represented in Fig. 5.14,
we noticed that the communication bandwidths between *DCT*, *Quantization*, and
Huffman modules are very high (640 bits) compared with those found between the
different other modules of the application (8, 24, and 32 bits). This bandwidth gap
will cause unbalanced traffic distribution especially when implemented on hardware,
since we will increase the link size in addition to the size and number of flits in the
packet format, causing higher latency and thermal power problem. All these factors
will eventually decrease the overall performance of the system, instead of enhancing
it. For all the reasons previously stated, we will implement the first task graph represented in Fig. 5.13 and we randomly (for simplicity) map the tasks on 2D-ONoC
(2×4) and 3D-ONoC $(2 \times 2 \times 2)$ as shown in Figs. 5.15 and 5.16 respectively.

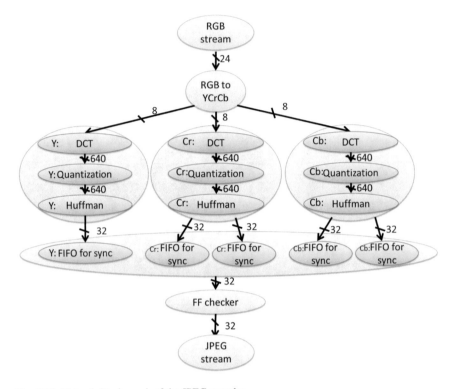

Fig. 5.14 Extended task graph of the JPEG encoder

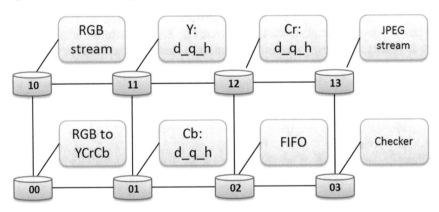

Fig. 5.15   JPEG encoder mapped on 2 × 4 2D-ONoC

5.2.7.2   Matrix Multiplication on 3D-ONoC

First, we assume that an $i \times k$ matrix A has i rows and k columns, where A_{ik} is an element of A at the i-th row and k-th column. As it is demonstrated in Fig. 5.17, an $i \times k$ matrix A can be multiplied by a $k \times j$ matrix B to obtain an $i \times j$ matrix R. Figure 5.18 presents how the matrix R can be obtained according to Formula 5.1:

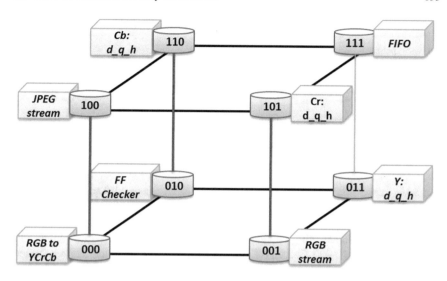

Fig. 5.16 JPEG encoder mapped on $2 \times 2 \times 2$ 3D-ONoC

$$\begin{pmatrix} A11 & \cdots & A1k \\ \vdots & \ddots & \vdots \\ Ai1 & \cdots & Aik \end{pmatrix} X \begin{pmatrix} B11 & \cdots & B1j \\ \vdots & \ddots & \vdots \\ Bk1 & \cdots & Bkj \end{pmatrix} = \begin{pmatrix} R11 & \cdots & R1j \\ \vdots & \ddots & \vdots \\ Ri1 & \cdots & Rij \end{pmatrix}$$

Fig. 5.17 Matrix multiplication example: The multiplication of an $i \times k$ matrix A by a $k \times j$ matrix B results in an $i \times j$ matrix R

$$R_{i,j} = \sum_{n=0}^{k-1} A_{i,n}.B_{n,k}. \tag{5.1}$$

When implemented onto 3D-ONoC, and for seeking of convenience or without loss of generality, we can assume that all the matrices are square and having $n \times n$ size. In 3D-ONoC, each element of the three matrices is assigned to a computation module which is connected to one router. As a result the number of routers connected to the network is the sum of all the elements of three matrices, which is equal to $3n^2$. Each element of the matrix B receives n flits from n different elements of the matrix A in order to make the multiplication. Then, each element of the matrix B sends n flits to n different elements of the matrix R where all the received values are summed and then the final resulted value is outputted. In total $2n^3$ flits travel the network for a nxn square matrix multiplication.

As we previously stated at the beginning of this section, we want to evaluate the number of hops traversed by all the flits generated by the matrix application. For this matter, we define

$$3D\_Hops_i = |x\_dest_i - x\_src_i| + |y\_dest_i - y\_src_i| + |z\_dest_i - z\_src_i|, \tag{5.2}$$

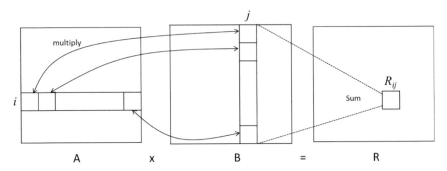

Fig. 5.18 Simple example demonstrating the matrix multiplication calculation

where $3D\_Hops_i$ is the number of hops consumed for one single flit $i \in \{0, 1, 2, ..., 2n^3 - 1\}$ (the set of all flits), traveling from one-source node (where the address is defined by x_dest, y_dest and z_dest) to its destination node (x_src, y_src and z_src). As a result, we can say that the number of hops consumed by an nxn square matrix multiplication can be defined by

$$3D\_Total\_Hops = \sum_{k=0}^{2n^3-1} 3D\_Hops_k. \tag{5.3}$$

According to Formulas 5.2 and 5.3, the number of hops for 2D-ONoC can be then extracted and defined as shown in Formulas 5.4 and 5.5:

$$2D\_Hops_i = |x\_dest_i - x\_src_i| + |y\_dest_i - y\_src_i| \tag{5.4}$$

$$2D\_Total\_Hops = \sum_{k=0}^{2n^3-1} 2D\_Hops_k. \tag{5.5}$$

For the evaluation, we took the case of 3×3, 4×4 and finally a 6×6 matrix multiplication. For each one of these three cases, two mapping approaches have been taken into consideration. For instance, we take the example of 3×3 matrix multiplication. We randomly mapped the elements of the three matrices into 2D-ONoC (3×9) and 3D-ONoC ($3 \times 3 \times 3$) using an optimistic mapping approach as presented in Fig. 5.19a. In this mapping we tried to make the communication distance as close as possible, in order to reduce the number of hops which eventually will lead to decrease the latency. Figure 5.19b, on the other hand, illustrates a pessimistic task mapping approach. The second approach tries to increase the communication path of the different flits traversing the network.

In order to obtain an easier and more accurate evaluation, both of 3D-ONCs are implemented in Verilog HDL. We evaluated and compared the hardware complexity in terms of area, power consumption, (static and dynamic) and clock frequency,

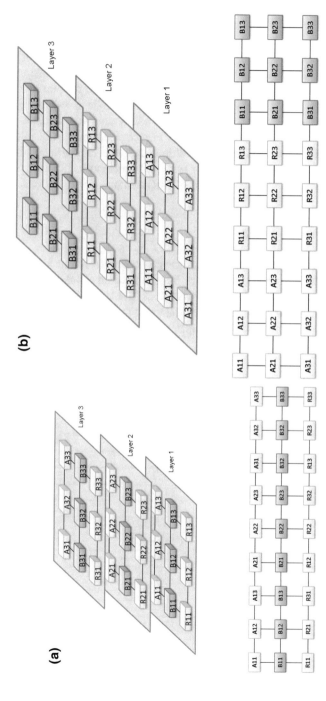

Fig. 5.19 3×3 matrix multiplication using **a** optimistic and **b** pessimistic mapping approaches

and also the performance in terms of execution time and the number of hops, and also we counted the number of *stop-signal* generated from our *Stall-Go* flow control mechanism. All the evaluation results obtained for 3D-ONoC are then compared to 2D-ONoC system.

We chose the Stratix III FPGA as a target device and then the synthesis was done by the Quartus II software, where both are provided by Altera. We used *Power-Play Power Analyzer* tool in Quartus II in order to evaluate the power consumption generated. This design approach results in more accurate speed, area, and power consumption evaluation. The use of FPGA is a very convenient choice for our design, thanks to its simplicity and the ability of reconfigurability. In addition to that, it provides faster simulation than the traditional software emulation while maintaining a cheaper cost than implementing with real processors. Table 5.1 presents the parameters used for the synthesis of 3D-ONoC design.

5.2.7.3 Evaluation Results

The goal of this section is to provide a hardware evaluation for the 3D-ONoC including area, power consumption, and clock frequency when simulated with both JPEG encoder and matrix multiplication applications. Table 5.2 illustrates the hardware evaluation results obtained. The results show that the logic utilization of 3D-ONoC is increased by an average of 37% compared to the 2D design. The increased number

Table 5.1 Simulation parameters

Parameters		2D-ONoC	3D-ONoC
Network size (Mesh)	JPEG	2×4	$2 \times 2 \times 2$
	Matrix (3×3)	3×9	$3 \times 3 \times 3$
	Matrix (4×4)	6×8	$4 \times 4 \times 3$
	Matrix (6×6)	9×12	$6 \times 6 \times 3$
Packet size	JPEG	3 flits	3 flits
	Matrix	1 flit	1 flit
Flit size	JPEG	30 bits	33 bits
	Matrix	35 bits	30 bits
Header size	JPEG	12 bits	17 bits
	Matrix	14 bits	17 bits
Payload size	JPEG	16 bits	16 bits
	Matrix	21 bits	21 bits
Buffer depth		4	4
Switching		Wormhole-like	Wormhole-like
Flow control		Stall-go	Stall-go
Scheduling		Matrix-arbiter	Matrix-arbiter
Routing		LA-XY	LA-XYZ
Target device		Altera stratix III	Altera stratix III

Table 5.2 3D-ONoC hardware complexity compared with 2D-ONoC

Apps	Area (ALUTs)		Power (mW)							F (MHz)	
	2D	3D	2D			3D				2D	3D
			S	D	Total	S	D	Total			
JPEG	28.401	30.382	811.63	4.27	815.9	769.13	4.01	773.14		193.8	160.72
M3 × 3	18.012	30.954	969.84	332	1301.84	1032.14	260	1292.14		158.73	130.01
M4 × 4	36.393	61.157	1073.52	495.2	1568.72	1055.65	410	1452.65		146.56	101.41
M6 × 6	89.576	144.987	1113.29	580	1693.29	1051.06	450.2	1501.26		98.85	98.1

of ALUTs can be explained by the fact that the 3D-ONoC router has two additional ports and a larger crossbar than 2D-ONoC. The additional number of ports incurs additional buffers, which is costly in terms of area.

In terms of clock speed, 3D-ONoC underperforms the 2D-ONoC architecture by 16% on average due to the increased hardware complexity. While the power static consumption is increased with 3D-ONoC with almost 14% for the same additional hardware reasons, the dynamic power on the other hand is decreased in average of 16% while executing JPEG and the two mapping approaches for each of the three matrix multiplications. As a conclusion, the total power consumption is decreased by nearly 1.4%. Many factors affect the dynamic power in FPGA, such as capacitance charging, supply voltage, and clock frequency. Since the first two factors are the same for both 3D- and 2D-ONoC designs, and only the clock frequency is different between them, we can say that the reduction of the clock frequency had an impact on the reduction of the dynamic power. Besides that the clock frequency reduction, we believe that the reduction of a number of hops (that will be explained in the next section) also plays an important role in the reduction of dynamic power. In fact, when the number of hops is reduced, it means that the flit has less hops, shorter path which eventually means less buffering, routing, and scheduling. All these factors lead to reduce the dynamic power when using 3D-ONoC when compared with 2D system.

5.2.7.4   Performance Analysis Evaluation

For the performance evaluation, we run each of the four applications. Then we evaluated the execution time, the number of hops, and the number of *stop-signal* of each one of them after verifying the correctness of the resulted data. Starting with the execution time, we run each of the four applications on 3D-ONoC and 2D-ONoC. Figure 5.20 demonstrates the execution time results. Taking a closer look at the JPEG application results, we may see that there is a slight improvement of 1.4% with 3D-ONoC when compared with the 2D architecture. This slight improvement can be explained by many reasons. First, JPEG is a small application which we could map into only eight nodes. That is a quiet small number to exploit the benefits of a 3D-NoC. Second, when observing the task graph of JPEG (previously shown in Fig. 5.13), JPEG has indeed some tasks working in parallel(*Y:d-q-h*, *Cb:d-q-h* and *Cr:d-q-h*), but at the same time we can see that *FIFO* module is dependent on those three tasks. Another reason is that the JPEG computation modules involve heavy computation. This leads to decrease the clock frequency of the entire system in a very inconvenient way for 3D-ONoC. The performance of 3D-ONoC is then hidden and cannot be taken advantage of. All of those reasons have an important impact on the performance of the 3D-ONoC. JPEG might be a very appropriate application to show the out performance of NoC over the traditional interconnect systems (such as bus-based system or P2P), but when we talk about 3D-ONoC that is targeted for hundreds of cores which is dedicated to a large number of cores with higher parallelism tasks.

Fig. 5.20 Execution time
comparison between 3D- and
2D-ONoC

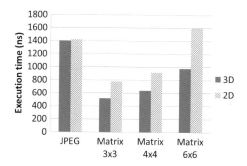

On the other part, when evaluated with the matrix multiplication application, 3D-ONoC shows a greater performance and decreases the execution time for about 35, 33, and 41% for each of 3×3, 4×4 and 6×6 matrix, respectively. In total 3D-ONoC reduces the execution time for one single matrix multiplication to up to 36% when compared with 2D-ONoC. As we stated previously, due to the fact that the matrix multiplication has a larger data array, higher number of parallel tasks with less dependency between them, matrix multiplication shows greater performance than JPEG. While the JPEG is mapped onto eight nodes only, the matrix multiplication can reach the 108 nodes for the 6×6 matrix size. These factors are very suitable to show the performance enhancement when adopting 3D-ONoC. This enhancement can be related to the reduction of a number of hops that offers 3D-ONoC.

Listing 5.12 Verilog-HDL code for hops number count

```
 1        for (i=1;i<=3;i=i+1) begin
 2          for (j=1;j<=3;j=j+1) begin
 3            for (k=1;k<=3;k=k+1) begin
 4              #200000
 5    //  **************Hop count from A to B**************
 6              if ((A_adress [i][j][2:0])>(B_adress [j][k
                  ][2:0]))
 7              Total_hops= Total_hops+ ((A_adress [i][j
                  ][2:0])-(B_adress [j][k][2:0]));
 8              else
 9              Total_hops= Total_hops+ ((B_adress [j][k
                  ][2:0])-(A_adress [i][j][2:0]));

11              if ((A_adress [i][j][5:3])>(B_adress [j][k
                  ][5:3]))
12              Total_hops= Total_hops+ ((A_adress [i][j
                  ][5:3])-(B_adress [j][k][5:3]));
13              else
14              Total_hops= Total_hops+ ((B_adress [j][k
                  ][5:3])-(A_adress [i][j][5:3]));

16              if ((A_adress [i][j][8:6])>(B_adress [j][k
                  ][8:6]))
17              Total_hops= Total_hops+ ((A_adress [i][j
                  ][8:6])-(B_adress [j][k][8:6]));
18              else
19              Total_hops= Total_hops+ ((B_adress [j][k
                  ][8:6])-(A_adress [i][j][8:6]));

    //  **************Hop count from B to R**************
```

```
22          if ((B_adress [i][j][2:0])>(R_adress [k][j
                ][2:0]))
23          Total_hops= Total_hops+ ((B_adress [i][j
                ][2:0])-(R_adress [k][j][2:0]));
24          else
25          Total_hops= Total_hops+ ((R_adress [k][j
                ][2:0])-(B_adress [i][j][2:0]));
26
27          if ((B_adress [i][j][5:3])>(R_adress [k][j
                ][5:3]))
28          Total_hops= Total_hops+ ((B_adress [i][j
                ][5:3])-(R_adress [k][j][5:3]));
29          else
30          Total_hops= Total_hops+ ((R_adress [k][j
                ][5:3])-(B_adress [i][j][5:3]));
31
32          if ((B_adress [i][j][8:6])>(R_adress [k][j
                ][8:6]))
33          Total_hops= Total_hops+ ((B_adress [i][j
                ][8:6])-(R_adress [k][j][8:6]));
34          else
35          Total_hops= Total_hops+ ((R_adress [k][j
                ][8:6])-(B_adress [i][j][8:6]));
36      end
37    end
38  end
```

Figures 5.21, 5.22, and 5.23 show the variation of the number of hops between 3D-ONoC and 2D-ONoC with 3×3, 4×4 and 6×6 matrix multiplications using pessimistic and optimistic mapping. The number of hops can be calculated using the Verilog code depicted in Code 5.12. This portion of code is added to the test bench that performs the calculation. When we analyze this figure, we may see that 3D-ONoC reduces the number of hops compared with the 2D system with an average percentage of 42, 31, and 47% 3×3, 4×4 and 6×6 matrices, respectively, having a total number of hops reduction of 40% over the 2D architecture. This can significantly reduce the execution time, since flits have fewer hops to traverse to reach their destination.

Another reason contributing to the performance of 3D-ONoC is the reduction of the traffic congestion. This can be seen by observing the *Stall-Go* flow control and the number of *stop-signal* generated by each matrix multiplication. To execute this calculation, we added a small portion of code (Code 5.13) at the end of the 3D-ONoC module, which uses the *net-stop-out* signal issued from the flow control and calculates the total stall count.

Fig. 5.21 Average number of hops comparison for both pessimistic and optimistic mappings on 3×3 network size

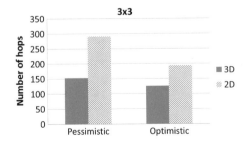

Fig. 5.22 Average number
of hops comparison for both
pessimistic and optimistic
mappings on 4 × 4 network
size

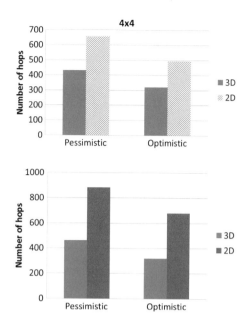

Fig. 5.23 Average number
of hops comparison for both
pessimistic and optimistic
mappings on 6 × 6 network
size

Listing 5.13 Verilog-HDL code defining for stall count

```
1   // 3D-ONoC top module: network.v
2
3   ...
4   ...
5   always @(reset) begin
6   if (reset) count <=    0;
7   end
8
9
10  always      @(net_stop_out)
11     begin  : stop
12            for (j=0;j<Y_WIDTH;j=j+1)begin
13               for (k=0;k<X_WIDTH;k=k+1)begin
14                  for (l=0;l<NOUT;l=l+1)begin
15                     if (net_stop_out[k][j][l]) count = count+1;
16                  end
17               end
18            end
19  end
```

As a matter of fact when observing Fig. 5.24, we can see that the stall count increase
linearly when we increase the matrix which is related to the number of flits traveling
the network. Even 3D-ONoC can reach up to 77% of stall count reduction over the
2D design with 6 × 6 matrix multiplication; the stall count impact cannot be clearly
seen with 3 × 3 and 4 × 4 calculation. This can simply be explained by the fact that
we are calculating a single matrix multiplication which generates only 54 and 128
flits for 3 × 3 and 4 × 4 matrix size, respectively. This small number of flits was not
enough to cause any traffic congestions in 3D-ONoC. For that reason, we decide
to extend the evaluation to calculate not only one matrix multiplication but also to
calculate 2, 3, and 4 different matrices at the same. This aims to increase the number

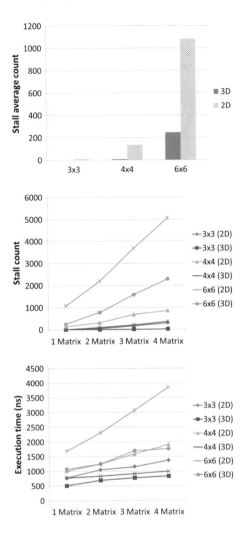

Fig. 5.24 Stall average count comparison between 3D- and 2D-ONoC

Fig. 5.25 Stall average count comparison between 3D- and 2D-ONoC with different traffic loads

Fig. 5.26 Execution time comparison between 3D- and 2D-ONoC with different traffic loads

of flits traveling the network at the same time to cause congestion. Then we evaluate again the average stall count.

Figure 5.25, depicts the average stall count of both 3D- and 2D-ONoC when implemented with 1, 2, 3, and 4 matrix multiplications. When analyzing this figure, the stall count has been dramatically decreased to 94, 67, and 59% in average for 3×3, 4×4, and 6×6 matrix multiplication, respectively. In total 3D-ONoC reduces the stall count to up to 74%. After calculating the stall number, we want to see the impact of increasing the traffic congestion on the execution time. So evaluate again the execution time of each matrix size when performing 1, 2, 3, and 4 matrix multiplications. The result obtained is shown in Fig. 5.26, which reduces the execution time to 36, 39, and 47% for 3×3, 4×4, and 6×6 matrix multiplication, respectively.

Then improving the total execution time reduction from 36% is obtained in the first experience with one matrix multiplication, to more than 41% when evaluated with heavier traffic load.

As the results mentioned above, 3D-ONoC take advantage of its ability to reduce the number of hops to enhance the performance. In addition, since 3D-ONoC router has two additional input–output ports, flits traveling the network have better routing choices which eventually will decrease the congestion that can be caused when using 2D-ONoC, having an important impact on the overall performance of the system. Not forget to mention, this will improve the traffic balance along the whole network which plays a very crucial role in the thermal power dissipated from the design.

5.2.8 Conclusion

Future applications are getting more and more complex, demanding a good architecture to ensure a sufficient bandwidth for any transaction between memories and cores as well as communication between different cores on the same chip. 2D-NoC architecture is efficient for medium-scale multicore SoC systems. However, soon it will not be probably a good candidate for large-scale heterogeneous many-core systems consisting of more than a thousand cores.

With the emergence of 3D integration technologies, a new opportunity emerges for chip architects by porting the 2D-NoC to the third dimension. In 3D integration technologies, multiple layers of active devices are stacked above each other and vertically interconnected using through-silicon via (TSV). As compared to 2D-IC designs, 3D-ICs allow for performance enhancements even in the absence of scaling because of the reduced interconnect lengths. In addition, package density is increased, power consumption is reduced, and the system is more immune to noise.

5.3 Photonic On-Chip Interconnect for High-Bandwidth Multicore SoCs

Photonic Network-on-Chip (PNoC) [43–47] is a novel concept enabling ultra-high communication throughput in the terabits per second range, low-power, and low-communication latency. When powered with a wavelength division multiplexing (WDM) scheme, multiple parallel optical streams of data are concurrently transferred through a single on-chip waveguide. This contrasts with the Electronic Networks-on-Chip, which require a unique metal wire per bit stream.

The key to saving power in PNoC systems comes from the fact that once a photonic path is established, the optical data are transmitted in an end-to-end fashion without the need for buffering, repeating, or regenerating. This is different from ENoCs, where messages are buffered, regenerated, and then transmitted on the inter-router links several times en route to their destination. Furthermore, photonic routers do not need to switch to every bit of the transmitted data like in electronic routers; optical routers switch on and off once per message, and their energy dissipation does not depend on the bit rate. This feature allows ultra-high bandwidth transmission while avoiding the power cost that is found in traditional ENoCs.

In a hybrid PNoC systems, the source node first issues a path configuration packet, which includes destination address information and other additional control information, via a copper-based electrical link. The configuration packet is routed via an Electric Control Network (ECN), reserving the photonic switches and channels along the path for the photonic message. When the photonic path reservation is completed, the source node returns an Acknowledgment (ACK) signal. When the ACK signal is received and processed by the source node, the optical data transmission starts. At the end of the transmission, all reserved photonic resources for the above data transmission are released.

The circuit-switched nature of such hybrid PNoCs directly affects the performance and power efficiency of on-chip communications. As observed in previously conducted study, the energy overhead of a hybrid PNoC system is mainly due to the electronic control modules, which consume the majority of the total power budget of a hybrid PNoC system. Moreover, the latency required for photonic path configuration is found to be much longer than the photonic data transfer itself.

While a single-layer configuration can provide low-loss waveguides and high-performance photonic devices, it suffers from limited integration density due to waveguide crossing and limited real estate. A way to go beyond this limitation is to monolithically stack multiple photonic layers above Si as multilayered electrical interconnections realized in modern electronic circuits [48, 49]. Figure 5.27 shows a high-level view of a three-dimensional PNoC (PHENIC) implemented with one electrical control layer and several photonic communication layers [50].

Fault tolerance is crucial when considering mission critical applications where the system must correctly function even when something goes wrong. One such application is that of space travel, where repair or replacement is not a possible option, and billions of dollars would be wasted.

5.3.1 Photonic Communication Building Blocks

The main components of an PNoC include a laser source, which generates phase coherent and equally spaced wavelengths, waveguides, which is used as a transmission medium, and modulators and photodetectors, which convert electrical digital data to and from photonic signals [51, 52]. Figure 5.28 shows a typical on-chip optical link that uses an external laser as a light source. It is expected that the laser source could produce up to 64 wavelengths per waveguide for a Dense Wavelength Division Multiplexing (DWDM) network.

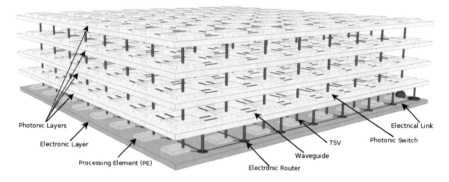

Fig. 5.27 3D-Stacked photonic network-on-chip architecture

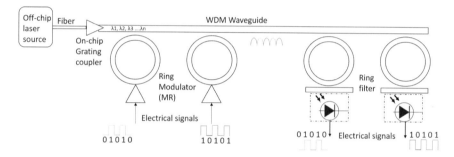

Fig. 5.28 Photonic link architecture

Laser Source: Since there are no available high-speed, electrically driven, on-chip monolithic laser light, PHENIC system features an off-chip laser source, such as VCSEL (Vertically Cavity Surface Emitting Laser). As indicated in Fig. 5.28, the off-chip laser source provides light to the modulator(s), which transduces electrical information into a modulated optical signals. Then, when the lights enter the chip, optical splitters and waveguides route it to the different modulators used for data transmission.

Modulators: Before optical messages are transmitted, the electrical messages from each IP core should be converted into optical form. PHENIC implements at each node a *gateway*: (Fig. 5.29) serving as a photonic network interface and based on silicon optical modulators and SiGe photodetectors. To reduce conversions time, modulators should be small (i.e., the circular-shaped 10 μm ring modulator [53]) and fast. The performance of a typical modulator is dependent on the on-to-off light intensity ratio [54], which depends on the electrical input signal strength. A higher extinction ratio is better and required for fast and accurate signal detection. Works in [53, 54] reported that an extinction ratio greater than 10 dB is acceptable and enough to enable proper signal detection without causing communication errors.

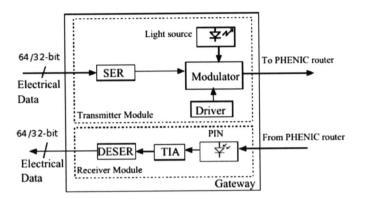

Fig. 5.29 Gateway organization

Waveguide: The waveguides provide the physical interconnection between all sources and destinations and enable connectivity between all photonic devices in PHENIC systems. The transmitter demultiplexes the light into appropriate wavelength channels and then modulates each of the channels with a digital data stream generated by the electronic component to be interconnected. Finally, photonic signals are routed to various PEs via routers and waveguides.

We have to note here that the *refractive index* [54] of the waveguide material has a big impact on the bandwidth, latency, and area of an optical interconnect. A waveguide typically has a width of $0.3\,\mu m$ [55]. Once the photonic signals are received by the destination node (receiver), the signals must be converted back to electrical form. Also, since PHENIC simultaneously transmits different wavelengths per bidirectional waveguides, a wave selective filter for each received wavelength is needed at the destination node.

Microring Resonator: The main element of a silicon-photonic NoC system is the microring resonator (MR). MRs are capable of effectively guiding an optical signal by carefully choosing their dimensions and positions along the path. Optical signals couple into ring resonators at specific regularly spaced wavelengths in the optical spectrum, called resonant modes [56].

5.3.2 Design Challenges

The photonic domain is immune to transient faults caused by radiation [57], but is still susceptible to process variation (PV) and thermal variations (TV) as well as aging. The aging typically occurs faster in active components as well as elements that have high TV [28]. In the optical domain, the faults can occur in MRs, waveguides, routers, etc. Active components, such as MRs, have higher failure rates than passive components, e.g., waveguides [28]. A single MR failure can cause messages to be misdelivered or lost, which results in bandwidth loss or even complete failure of the whole system. Together, fabrication-induced PV and TV effects present enormous performance and reliability concerns. TV causes a microring to respond to a different wavelength than intended. This can take the form of a passband shift in the MRs. When an MR heats up, it expands, changing its radius, and therefore shifting the wavelengths which it uses to the right [58]. As reported in [59], a change of as little as $1°$ C can shift the resonance wavelength of a microring by as much as 0.1 nm. This is not permanent and will return when the temperature returns to normal. Therefore, systems' temperature must be kept at a reasonable value in order for the MRs to resonate correctly. This is challenging, especially in a large complex computing system, which uses thousands of these components. Trimming technique [60] is generally used to dynamically modify the resonance frequency of a microring to overcome both thermal drift and fabrication inaccuracy. This technique can be accomplished by dynamically increasing the current in the $n+$ region or by heating the ring [60–62].

PV is the variations of critical physical dimensions, e.g., thickness of wafer and width of waveguides, and also affects the resonant wavelengths of MRs. This means that not all fabricated MRs can be used due to PV. As a result, network nodes that do not have all working MRs would lose some or all of wavelengths/bandwidths in communication [63]. To solve this problem, Xu et al. [64] proposed a method of flexible wavelength assignment. Because the networks are already built with excess detectors or modulators for each message, the node with the excess components can compensate and rematch the components which have been affected by PV.

Over time, all silicon-based ICs wear down. We refer to this phenomenon as *aging*. Some of the aging effects only apply to the active components, because of their electrical subcomponents [65], such as the MRs, while other aging affects all parts, even the waveguides.

Recent PNoCs researches (i.e., network topology, router micro-architecture design, and performance and power optimization and analysis) have resulted in several architectures capable of transmitting at a high data bandwidth and low energy dissipation [43–47]. In [48], we proposed an energy-efficient and high-throughput hybrid silicon-photonic network-on-chip based on a smart contention-aware path-configuration algorithm and an energy-efficient non-blocking optical switch to further exploit the low energy proprieties of the PNoC systems. However, little attention has been given to the aspect of fault tolerance and reliability along the photonic interconnects.

This chapter presents a fault-tolerant PNoC architecture. The system is based on a fault-tolerant path-configuration and routing algorithm, a microring fault-resilient photonic router, and uses minimal redundancy to assure accuracy of the packet transmission even after faulty MRs are detected.

5.3.3 Fault Models

It is worth noting that the light is not sensitive to radiation or electromagnetic fields, the signals which control the optical network can be sensitive to it. The following is a list of actual possible causes that can contribute to the failure of an optical device.

5.3.3.1 PNoC Signal Strength

Typical NoCs are defined by their power consumption, delay, and throughput. PNoCs also have to consider the signal-to-noise ratio at the receiving end. Because they do not buffer and retransmit, the signal gets weaker based on how many hops it jumps. This does not significantly affect the power the network consumes, but it can lead to a higher sensitivity to noise.

5.3.3.2 Electrostatic Discharge

While the waveguides are not electrically conductive, the switches and photodetectors are. This means that they are sensitive to high currents. One thing which can ruin an IC is electrostatic discharge (ESD). This is when a current enters through the I/O pins of the control circuit, or it can be caused by an extremely strong magnetic field. This all results in the aforementioned extreme current, and this current causes severe damage to the silicon in the components. Possible points of damage are the dielectric, the PN junctions, and any wiring connecting to the controllers. Because of the scaling, the causing phenomena have become harder to control [66]. This can be prevented by proper packaging to the IC providing ESD protection at the pins.

5.3.3.3 Noise

This is one of the unique things that we categorize as a cause for a fault. The reason is because the noise can be caused simply by poorly matched wavelengths. It can also be caused by creating a path that is too long, or a path that crosses too many intersections. These paths tend to be caused by rerouting or non-minimalistic routing, but other factors can contribute and cause more noise. The most common factors are listed in the following subsections.

5.3.3.4 Aging

Over time, all silicon-based ICs wear down. Some of the aging effects only apply to the active components, because of their electrical subcomponents, while other aging affects the optical properties of the components.

Electromigration—This mainly affects the wires which control the ring resonators. It does not affect the waveguides in any way. It originally causes a delay in the wire, and can eventually lead to an open, or to a short to a nearby wire. It achieves this by thinning out the thinnest portion of the wire due to the higher current density at the bottleneck [67].

Laser Degradation—After the lasers have been on for several hundred hours, they start to show signs of degradation. This shows in the form of either missing wavelengths, which can cause a channel fault, or general weakening of the original laser signal. In each of these cases, it does not become a true problem until the signal falls to a level where the worst-case scenario's signal-to-noise ratio is too weak to receive an understandable signal [68].

Photodetector Degradation—Various studies have been done for different types of photodetectors showing that they degrade overtime, particularly from being exposed to thermal conditions or UV light. It is reasonable to assume that no matter what material photodetectors are made out of, they all seem to be vulnerable to degradation due to thermal variation, which is present in all networks [28, 65].

A lot of work has been done to combat the effects of aging. Some examples are Agarwal [69], Keane [67], and Kim [70]. These are mainly focused on the electrical side, but the fact that these do exist shows the hope for a future where optical aging can be researched and prevented. Many parameters such as the wavelengths and laser strength can possibly be modified throughout the life of a chip to counteract the aging effects in a similar manner to what Mintarno does for Electrical networks [71].

5.3.3.5 Process Variability

This can affect both the active and inactive components of the optical network. The variability accounts for material impurities, doping concentrations, and size and geometries of structures [72]. One single dimple in a particular point in the coupling region of a ring resonator can greatly affect the coupling properties and thus cause problems for the switch, or maybe just the channel. A poor geometry can also cause a certain component to be more sensitive to aging or ESD. Obviously, if a variation gets bad enough, an entire link can be rendered useless. This would be considered an early permanent fault and should be detected before a device is released. The impurities in a waveguide can cause such a block, or cause there to be a change in the reflectivity of the material, and that causes a higher amount of insertion loss, resulting in a lower signal-to-noise ratio. Other similar chains of events can occur from bad doping of the photodetectors. Minimizing this process variability can greatly increase the reliability of the system, even without implementing fancier and area or energy heavy redundancies. The unfortunate truth is that with recent advances in scaling, the variability continues to increase [73, 74].

5.3.3.6 Temperature Variation

For electrical components, temperature variation can cause changes in properties such as resistivity and cause more power consumption or delay, but in the optical domain, it is quite different. Ring resonators are tuned by heating up the ring, causing them to expand, which changes their passband wavelength. If the chip heats up to a point beyond the tuning, then certain channels just disappear as a whole. The increase in temperature also causes the photodetectors to degrade as mentioned in the previous section. These temperature variations also tend to speed up other forms of aging as well.

5.3.4 Fault-Tolerant Photonic Network-on-Chip Architecture

The Fault-tolerant Photonic Network-on-Chip (FT-PHENIC) system, shown in Fig. 5.30, is a mesh-based topology and uses minimal redundancy to assure accuracy of the packet transmission even after faulty MRs are detected [75]. The system uses Stall-Go mechanism for flow control, and a matrix arbiter as a scheduling technique [2, 19, 76]. FT-PHENIC is also based on a microring fault-resilient photonic router (MRPR) and an adaptive path-configuration and routing algorithm [50, 77].

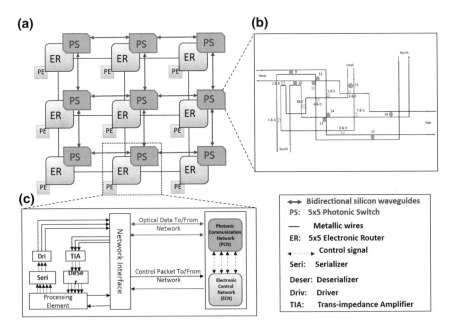

Fig. 5.30 FT-PHENIC system architecture. **a** 3 × 3 mesh-based system, **b** 5 × 5 non-blocking photonic switch, **c** Unified tile including PE, NI, and control modules

As illustrated in Fig. 5.30, the proposed system consists of a Photonic Communication Network (PCN), used for data communication, and an Electronic Control Network (ECN), used for path configuration and routing. Each PE (Processing Element) is connected to a local electrical router and also connected to the corresponding gateway (modulator/detector) in the PCN [48]. Messages generated by the PEs are separated into control signals and payload signals. Control signals are routed in the ECN and used for path configuration and routing. The payloads are converted into optical data and transmitted on the PCN.

5.3.4.1 Microring Fault-Resilient Photonic Router

The block diagram of the Microring Fault-resilient Photonic Router (MRPR) is shown in Fig. 5.31. It consists of a non-blocking fault-tolerant photonic switch (Fig. 5.31a) and a light-weight control router (Fig. 5.31b). Redundant MRs are carefully placed at special locations on the switch to assure fault tolerance even if one of the MRs on the backup path has a fault. The backup route for the NEWS (North–east–west–south) directions is to actually use the waveguide connected to the core ports as a master backup; therefore, the redundant MRs are all chosen at the locations which connect the NSEW ports to the core.

For a majority of faults, the design of the switch allows for an alternate, slightly less power efficient route. In fact, the backup route is less power efficient because the packets travel across more waveguide distance, go through more active MRs, and cross more waveguides. However, the switch still maintains all of its functionality. Because backup routes are only intended for use in the switches in which faults have occurred, the extra loss will have minimal effect on the message' signal strength across the whole network.

The MRPR was designed to require no MRs from east–west and north–south traffic. Since this kind of traffic accounts for a majority of the traffic of the PCN [50], such design will save on power and continue to function in the case of any MR fails. Assuming that a single location of redundant MRs does not fail altogether, the switch is able to maintain all functionality at slowed speeds.

Figure 5.32 shows a reconfiguration example of how MR 9 can be backed up by MRs 5, 15, and 1. Additionally, the MRs which connect parallel waveguides are replaced with racetracks [78]. This allows for a wider passband of light frequencies and makes them less sensitive to physical faults, such as reduced sensitivity to thermally caused passband shifting. Racetracks also have a larger Mean Time Between Failures (MTBF) [78].

The original form of MRPR switch is a five-port non-blocking switch, meaning that it allows for routing from any available port to any other available port. Once a fault is detected, the switch recovers, but there is a chance that it may turn into a blocking switch; however, it should be able to maintain all functionality as long as none of the redundant MRs fails. Because the redundant MRs lie dormant, they do not require much power other than the boost in signal strength required to compensate for the signal loss, caused by passing an inactive MR, which is minimal. As all rerouting

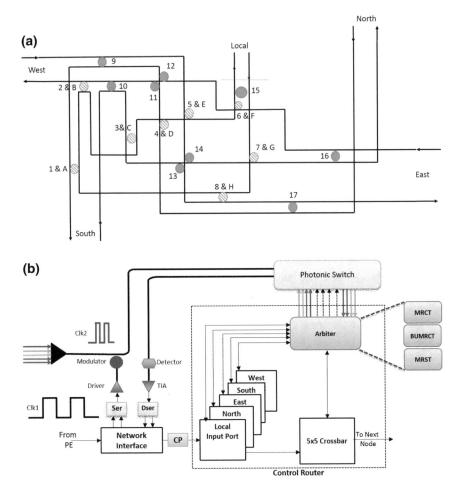

Fig. 5.31 Microring fault-resilient photonic router (MRPR): **a** Non-blocking fault-tolerant photonic switch, **b** Light-weight control router

in the switch occurs on the core waveguide, traffic certainly increases on this one waveguide as too many faults occur, which is why it should be treated as a node failure after a threshold of failed MRs is reached.

In addition to tolerating faults, MRPR is able to handle the *ACK* signals and the resulting regeneration process of the *tear-down* signal at each hop. To accomplish this goal, a hybrid switching policy is used: *Spacial-switching* for the data signals by manipulating the state of the broadband switching elements and a *Wavelength-selective switching* for the *tear-down* signals by using detectors and modulators. Moreover, since the *tear-down* signals should be checked and regenerated at each hop, it is crucial that their manipulation be automatic and not interfere with data signals, nor cause a blockage inside the switch. When the *tear-down* is generated

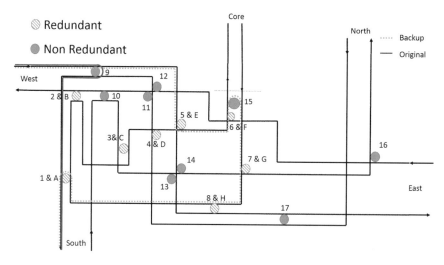

Fig. 5.32 Example of how a non-redundant MR's functionality can be mimicked by redundant ones

Table 5.3   Microring configuration for normal data transmission

Output/Input	Core	North	East	South	West
Core	–	4	6	3	5
North	7	–	16	None	14
East	8	17	–	13	None
South	1	None	12	–	9
West	2	11	None	10	–

at the source NI (Network Interface), it is first sent to the control router. Then, the *photonic switch controller* releases the corresponding MRs and generate another *tear-down* which is sent to the output-port modulator in the PCN where it continues its path in a hop-by-hop basis until it reaches its destination. At the destination node, the *tear-down* is detected in the input port and sent to the *photonic switch controller* in the corresponding electronic router. In this fashion, we can omit the overhead of an additional gateway which becomes significant when we increase the number of cores. Table 5.3 shows the MRs configuration for data transmission, where 16 MRs are used in a non-blocking fashion. Table 5.4 shows the backup paths for each transmission.

We use the first six wavelengths in the optical spectrum starting from 1550 nm, with a wavelength spacing equal to 0.8 nm to maintain a low cross-talk as reported in [79]. For the acknowledgment signals, we use the first five wavelengths in the optical spectrum starting from 1550 nm: four wavelengths for the *tear-down* signal where each one is dedicated to each port except the local one. In addition, a single wavelength is used for the *ACK*. The remaining available wavelengths are used for data transmission. The five wavelengths used to control the *ACK* and *tear-down*

Table 5.4 Microring backup configuration for data transmission

Output/Input	Core	North	East	South	West
Core	15	D	F	C	E
North	G	–	6,15,7	None	5,15,7
East	H	4,15,8	–	3,15,8	None
South	A	None	6,15,1	-	5,15,1
West	B	4,15,2	None	3,15,2	–

Table 5.5 Wavelength assignment for acknowledgment signal (Mod: Modulator, and Det: Photodetector)

	Core	North	East	South	West
Input	Mod_{λ_0}	Det_{λ_3}	Det_{λ_2}	Det_{λ_1}	Det_{λ_4}
Output	Det_{λ_0}	Mod_{λ_1}	Mod_{λ_4}	Mod_{λ_3}	Mod_{λ_2}

signals are notably constant regardless of the network size, in contrast with the fully optical where the number of wavelength used for control and arbitration grows with the network size. Thus, cutting these wavelengths from the available spectrum to be used for control would not degrade the system bandwidth. These five wavelengths will be negligible especially when DWDM is used providing up to 128 wavelengths per waveguide [80]. The wavelength assignment for each port is shown in Table 5.5.

Should the *tear-down* signals enter the switch, they need to be redirected to the corresponding electronic router. Since these signals are coming from different ports, and are modulated with different wavelengths, detectors capable of switching all of the four wavelengths are placed in front of the input ports to intercept the signals. The converted optical signal will be redirected to the electronic router to be processed. According to the included information, the corresponding MRs will be released. For the *ACK*, when the PSCP reaches the destination, 1-bit optical signal is modulated starting from the output port (i.e., opposite direction) and travels back to the source.

With this smart hybrid switching mechanism, we take advantage of the low-power consumption of the optical link by using optical pulses modulated with the adequate wavelength instead of propagating the acknowledgment signals in the ECN. Second, we take advantage of the WDM proprieties by separating the acknowledgment packets and the data signals and let them coexist in the same medium without interfering with each other. This contrasts with the electronic domain where these acknowledgment packets travel for several hops consequently blocking (preventing) the waiting cores from sending their PSCP packets. Finally, we are able to tolerate faults due to the arrangement of the MRs, and allowing for redundancy at critical locations.

As a primary comparison, we performed a study on the routers, and the loss that they would each have on average, and in their worst case. The results can be seen in Table 5.6. As expected, the Crux [81] performs the best, as its only design goal was to minimize loss and noise, sacrificing a lot of functionality. Values for the calculation were taken from various authors and can be seen in Table 5.7.

Table 5.6   Various switches and their estimated losses. AL: Average Loss, WL: Worst Loss

Router	Cros.	MRs	Termi.	AL (dB)	WL (dB)	WL (faulty) (dB)
Crossbar	25	25	10	1.12	1.60	∞
Crux	9	12	2	0.657	1.11	∞
PHENIC	27	18	0	1.315	1.615	∞
FT-PHENIC	19	16 + 9	0	0.965	1.115	2.215

Table 5.7   Insertion loss parameters for 22 nm process

Parameter	Value (dB)
Through ring loss	0.5 [81]
Pass by ring loss	0.005 [82]
Bending loss	0.005 [82]
Crossing loss	0.12 [81]
Terminator	0.01 [82]

5.3.4.2   Light-Weight Electronic Control Router

Figure 5.31b illustrates the control router architecture, which is based upon OASIS-NoC router [2, 76, 83, 84]. As shown in the above figure, the arbiter receives the detected *tear-down* from the above switch (colored arrows). According to the information encoded in this signal, the corresponding MRs are released and a new *tear-down* is generated for the next hop until it reaches its final destination and all MRs involved in this communication are released. The figure shows also the connection between the network interface (NI) and the local port, where a configuration packet (CP) is sent from the NI to the local port. The CP could be a setup packet or a path-blocked packet. The NI is connected also to the data switch (i.e., PCN). When the source node receives the ACK, the payload is processed by a serializer bank (if needed), a high-speed driver, and a modulator to convert the electrical signal into an optical one. At the source node, the optical data leaves the data switch and go through a detection step, a high-speed Trans-Impedance-Amplification step, and a deserialization step. At the end the NI's receiver receives the payload data with its original clock speed.

5.3.4.3   Fault-Tolerant Path-Configuration and Routing

The key feature of the Fault-tolerant Photonic Path-configuration algorithm (FTPP) is that it can handle faulty MRs within the photonic switches. When a fault occurs, the algorithm checks for the secondary MRs on the list and checks their status. The backup MR table can be very simple in the cases of a redundant MR failing, where it is simply replaced by its redundancy, or it can be slightly more complicated, as shown in Fig. 5.32.

The FTPP algorithm must meet certain requirements to work with the FT-PHENIC system. It should be also able to remove the dependency between the ECN and PCN which causes a significant latency overhead in conventional hybrid PNoC systems. In addition, the latency caused by the path blocking, which requires several cycles for the path dropping and the new path setup packet generation, is considerably decreased. Another key feature of the configuration algorithm is the efficiency of the ECN resources' utilization. By moving the acknowledgment signals to the upper layer, we can reduce the buffer depth to only two slots, since half of the network traffic is eliminated. This reduction is a key factor to design a light-weight router, highly optimized for latency and energy.

Figure 5.33a shows an example of a successful path setup process where all the necessary resources between a given source–destination pair are reserved. The corresponding pseudocode is given in the algorithm shown in Fig. 5.34.

Before optical data transmission, the source node issues a *Path-setup-Control-Packet* (PSCP) which is routed in the ECN and includes information about the destination and source addresses. In addition to the source and destination addresses, other information is included. For example, one bit is used for the packet-type field. This field can be "00" for a PSCP and "01" when this configuration packet is a path-blocked. Other information to ensure quality-of-service and fault tolerance, such as message ID, fault status, and error detection code, can also be included.

For each electrical router, the output port is calculated according to dimension-order routing [84]. Every time the PSCP progresses to the next router, the optical waveguides between the previous and current routers are reserved. Depending on the output port of the electrical router, the corresponding photonic router is configured by switching on/off one or more MRs using the MRs configuration table shown in Table 5.3.

In the example shown in Fig. 5.33a, the packet is entering the local input port attached to the Network Interface (NI) and requesting the east output port. According to Table 5.3, MR 8 is required and its availability is checked in the (Microring State Table) MRST. In this table, the MR's state is "00" (free and not faulty). Therefore, the switch controller reserves the MR and changes its states from "00" (free and not faulty) to "01" (not free and not faulty). After this successful reservation (hop based), the PSCP continues its path to the next hop and the same procedure is repeated until all necessary MRs are reserved for the complete path. This process is illustrated in *lines* 1–10 of the algorithm shown in Fig. 5.34.

In a case where the requested MRs at a given optical switch along the path are not available, blocking occurs. This can be seen in Fig. 5.33b where *MR* 5, which is necessary for the ejection to the local output port from the west input port, and is used by another communication. In this case, the *PSCP* is converted into a *path-blocked* packet (PB). The PB, then, travels back to the source node and releases the already reserved resources. The release is done by re-updating the corresponding entries in the MRST to "00" and by sending an electrical "OFF" signal to the corresponding MRs in the PCN. This process is illustrated in *lines* 11–15 of the algorithm shown in Fig. 5.34.

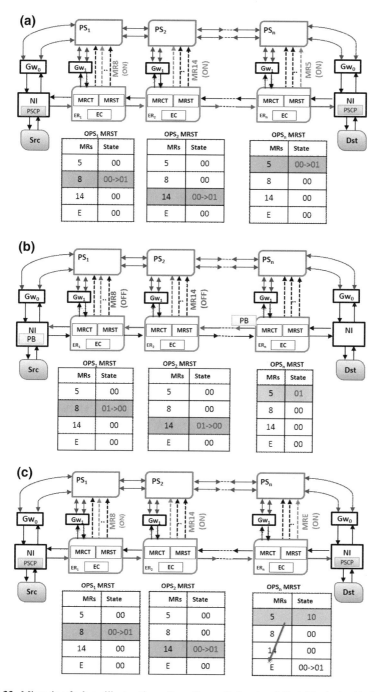

Fig. 5.33 Microring fault-resilient path configuration: **a** Path setup, **b** Path-blocked, **c** Faulty MR with recovery. GW_0: Gateway for data, GW_1: Gateway for acknowledgment signals, PS: photonic switch, MRCT: Microring Configuration Table, MRST: Microring State Table. 00 = Not faulty, Not blocked, 01 = Not faulty, Blocked, 10 = Faulty

```
    // Path Setup Control Packet for communication i, PSCPi
    // Path Blocked Packet for communication i, PBi
    Input: S_i, D_i
    // From ACK detector
    Input: Detc_ACKs
    // To ACK modulator
    Output: Mod_ACKs
    // From Teardown detector
    Input: TeardMod_i
    // To Teardown modulator
    Output: TeardMod_i
    // To Microring resonator
    Output: MRs_{j=0...n}
    // Buffer writing and routing computation stages
 1  initialization;
 2  while (Path-Setup-Control-Packet (PSCP) !=0) do
 3      DestAdd ← PSCPi;
 4      PortIn ← PSCPi;
 5      if (resource are available ) then                    /* check MRs state */
 6          if MR is not faulty then
 7              Grant_i ← Arbiter;
 8          else if Backup MR is Not Faulty then             /* check backup MRs state */
 9              GrantBackup_i ← Arbiter;
10          else                                             /* no possible path */
11              Blocked_i ← Arbiter;
12              FaultyNodeList ← Node;
13          end
14      else                                                 /* generate path blocked */
15          Blocked_i ← Arbiter;
16      end
17  end
    // Path blocked
18  initialization;
19  while ( PB !=0) do                                       /* Path blocked arrives */
20      if (MRsi state is reserved) then                     /* release reserved MRs */
21          release ← MRsi;
22  end
    // Generate ACK
23  initialization;
24  while (NI receiver ← PSCPi) do                           /* PSCP arrives to Dest */
25      if (PSCP arrives to NI) then                         /* generate ACK to Src  */
26          ACK_i ← To modulator ACK (λ0);
27  end
    // Receives ACK
28  initialization;
29  while (NI receiver ← ACK_i (λ0)) do                      /* ACK arrives to Src λ0 */
30      if (ACK arrives to the NIsender ) then               /* modulate the data    */
31          Data_i ← To Data's Modulator;
32  end
    // Identify and Generate Teardown_i
33  initialization;
34  while (From detector signal =Teardown_i with λi) do
35      findInport ← λi;            /* find In-port according to the wavelength */
36      free ← MRsi;                               /* Free involved MRs */
37      Teardwon_i ← To modulator λi;   /* generate new Tear-down according to λi */
38  end
```

Fig. 5.34 Fault-tolerant path-configuration algorithm

If a fault is encountered along the way, denoted by a state of "10", shown in Fig. 5.33c, then the switch attempts to use its backup route within the switch to maintain the intended port-to-port communication. This allows for recovery without requiring the whole system to change the route of a packet, and can save on costly retransmission and multiple attempts at setting up the path. Assuming that the backup path is being used for a recovery path, the algorithm proceeds with sending the standard path-blocked packet.

When the *PSCP* arrives successfully at the destination node, the NI modulates one-bit acknowledgment (ACK) signal to travel back to the source via the PCN. This can be seen in lines 16–20 of Algorithm 1. Upon the arrival of this *ACK* signal, the source node modulates the payload through the data modulators and sends it to the destination node via the PCN. Lines 21–25 of the algorithm shown in Fig. 5.34 depict this data/payload transfer phase. The last process of the proposed path-configuration algorithm is the *tear-down* step as shown in lines 26–31 of the algorithm shown in Fig. 5.34. When the entire payload is transmitted, it is necessary to release the reserved optical resources. This is handled by the source node which sends a *tear-down* packet to the destination after predetermined number of cycles depending on the source–destination addresses, transmission bandwidth, and message size.

The source's NI sends the electronic *tear-down* packet (TD) to the first electronic router ER_1. The Electronic Controller (EC) in this router indexes the MRCT with input–output ports information and determines the MRs that need to be released. As we can see in this figure, the state of MR 8, previously reserved in the path setup process, is reset to *Free* (state = "00") and electrical "OFF" signals are sent to the MR.

After the MRs are deactivated, a new optical tear-down signal is generated according to the used wavelength. It is sent through the PCN to the next hop where it is converted back to electrical and redirected to the EC in the corresponding electronic router to be processed. After this process, the MRs are released and a new optical tear-down signal is generated. This process is repeated until the *tear-down* reaches the destination and all optical resources are released. It is important to mention that the path setup and path-blocked processes of the proposed algorithm are very similar to the conventional ones [43, 44, 85–88]. The main difference is that the MRST in our proposal contains only two states: *Free* and *Active*. The MRs are set "ON" as soon as the PSCP succeeds to reserve them. In the conventional mechanisms, three states are necessary: *Free*, *Reserved*, and *Active*. When the PSCP finds the requested MRs *Free*, it updates their states in the MRST to *Reserved* without turning them "ON". When the complete path setup process is completed, the ACK signal travels back to the source node and sets the corresponding MRs "ON" by updating their states in the MRST to *Active*. With the proposed algorithm, some portions of the reserved path might be set "ON" and then "OFF" due to the unavailability of the resources. However, it enables the fast ACK transmission in the PCN.

In conventional path-configuration algorithms, the ACK and tear-down packets are transmitted in the ECN and have to go through all the buffering, routing computation, and arbitration stages. With the proposed algorithm, they are carried via the PCN. As a consequence, the ETE latency can be significantly reduced in addition to the dynamic energy saving that can be achieved. Additionally, conventional path-configuration algorithms do not check for faulty MRs. This will allow the system to tolerate more MR failures and take advantage of the fault-tolerant switch.

5.3.5   Evaluation

We evaluate the FT-PHENIC system using a modified version of PhoenixSim which is developed in the OMNeT++ simulation environment [82]. The simulator incorporates detailed physical models of basic photonic building blocks such as waveguides, modulators, photodetectors, and switches. Electronic energy performance is based on the ORION simulator [89]. We evaluate the bandwidth performance and energy consumption for 16, 64, and 256 cores systems (Table 5.8).

We compare the performance of the FT-PHENIC systems with the baseline PHENIC [48], and the system using the algorithm proposed by Xiang et al. [63]. Xiang's network was chosen over other typical systems [90–93], because it uses some form of fault tolerance, and most of their results would mimic the baseline PHENIC.

For the fault-related data, we disabled a certain number of MRs at random, and recorded the data. To get better results, we would run each system at each fault rate 10 times, and then averaged each test's total energy, average bandwidth, and average latency. Currently, the MR is disabled for the whole test, and thus models either a permanent or intermittent fault. Dealing with passband shift or temporary overheating of an MR is outside of the scope of this paper, beyond redundancy as a solution. The fault rates were chosen to span from 0 to 30% due to the fact that at this point, all of the tested networks were in deadlock (Table 5.9).

Table 5.8 Configuration parameters

Network configuration	Value
Process technology	32 nm
Number of tiles	256,64,16
Chip area (equally divided amongst tiles)	400 mm^2
Core frequency	2.5 GHz
Electronic control frequency	1 GHz
Power model	Orion 2.0
Buffer depth	2
Message size	2 kbytes
Simulation time	10 ms (25 10^8 cycles)

Table 5.9 Photonic communication network energy parameters

Network configuration	Value
Datarate (per wavelength)	2.5 GB/s
MRs dynamic energy	375 fJ/bit
MRs static energy	400 μ W
Modulators dynamic energy	25 fJ/bit
Modulators static energy	30 μ W
Photodetector energy	50 fJ/bit
MRs static thermal tuning	1 μ W/ring

5.3.5.1 Complexity Evaluation

The complexity evaluation considers the number of used rings and the resulting static thermal tuning. The number of used MRs is given by Eq. 5.6, where $Mod/Detc_{(ring)}$ is the number of rings required to modulate/detect the payload signal. $Switch_{(ring)}$ is the number of ring required for the photonic switch to route the optical data. Finally, the $ACKs(ring)$ is the number required to handle the acknowledgment signal:

$$Total_{(ring)} = Mod/Detc_{(ring)} + Switch_{(ring)} + ACKs(ring). \qquad (5.6)$$

Tables 5.10 and 5.11 show the comparison results for 64 and 256 cores systems, respectively. We can see that the optimized networks have the lowest number of rings. In fact, this kind of network is even more sensitive to MR faults as each MR is critical for the functionality of the node. In addition, with a minimal number of rings, the resulting insertion loss is lower than the fault-tolerant design. For the proposed FT-PHENIC system, it has additional rings used for acknowledgment signal, compared to the other networks, as well as for fault tolerance. This increase can reach 33% when compared to the optimized crossbar and PHENIC systems. We also observe the same behavior when evaluating the required static thermal tuning, which is required to maintain the functionality of the ring, under 20 K temperature with 1 μ W for each ring.

Table 5.10 MR requirement comparison results for 64 cores systems

	FT-PHENIC	PHENIC	Xiang
Mod/Detc	64	64	64
Switch	1152	1152	1600
ACKs	640	640	–
Redundant MRs	384	–	–
Total	2240	1856	1664
Sta. Power (mW)	44	37	33

Table 5.11 MRs requirement comparison results for 256-core systems

	FT-PHENIC	PHENIC	Xiang
Mod/Detc	256	256	256
Switch	4608	4608	6400
ACKs	2560	2560	–
Redundant MRs	1536	–	–
Total	8960	7424	6656
Sta. Power (mW)	179	149	133

5.3.5.2 Latency and Bandwidth Evaluation

Figure 5.35a, b shows the overall average latency and the average latency near the saturation region, respectively. We can see that for zero-load latency, all networks behave in the same way. Near saturation, PHENIC shows more flexibility and scalability in 256 cores when compared to the other networks. For the 64 cores configuration, the crossbar-based system slightly outperforms both PHENIC systems in terms of latency. This can be explained by the use of optical-to-electronic conversion of the *tear-down* which affects the overall latency of small networks.

The latency is heavily affected by the failure rate of MRs, and as the systems fail more, the latency increases until the whole system fails. This has a lot to do with failed path setup. Figure 5.36 shows the results of the latency test when adding in varying amounts of MR failures. The FT-PHENIC demonstrates its ability to withstand MR failures over all other systems.

For the achieved bandwidth, Fig. 5.37 shows that the bandwidth is increased by about 51% when compared to Xiang' system, for both 64 and 256 cores configurations. When compared to the crossbar, torus and PHENIC systems, we see that the four systems behave similarly. While the torus system has the capability of setting the path with less hop count, the FT-PHENIC system can achieve the same performance without the need for an extra network access which is required for the torus. This behavior is observed for 16, 64, and 256 core systems.

The latency increase caused by failed MRs will in turn cause the bandwidth to decrease. The effects of the failures on the bandwidth can be seen in Fig. 5.38. As with the latency, only FT-PHENIC and Xiang show any tolerance to faults, with FT-PHENIC outperforming Xiang.

5.3.5.3 Energy Evaluation

Figure 5.39 shows the total energy and the energy efficiency comparison results for 16, 64, and 256 cores systems. For the 256 cores configuration, the proposed system outperforms all other networks. This is illustrated by an improvement in terms of energy efficiency reaching 26% when compared the crossbar-based (non-blocking).

Fig. 5.35 Latency comparison results under random uniform traffic: **a** Overall Latency, **b** Latency near saturation

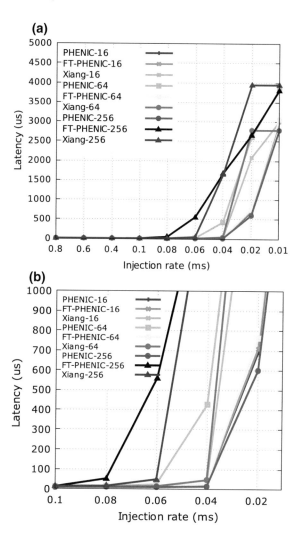

When compared to the torus-based architecture, FT-PHENIC improves the energy efficiency by upward of 70%. The torus-based architecture offers high bandwidth, thanks to the connection between edges leading to short communications. On the other hand, it comes at high energy cost. This can be explained by the fact that the additional input ports, required for the edge connections established in the torus-based system, incur increased area and consequently an energy overhead.

Figure 5.40 shows the total energy and energy efficiency of the systems when 4% of their MRs have failed. Some systems were not able to complete simulation, and so their energy is marked as 0 J, and an efficiency of 0 pJ/bit, just so the functioning ones remain visible. The extra energy comes from the extra run time. It is important

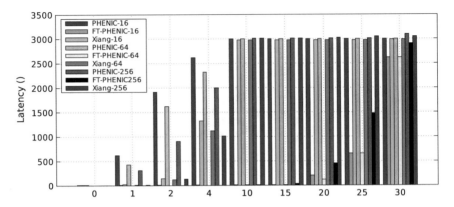

Fig. 5.36 Latency results of each system as faults are introduced

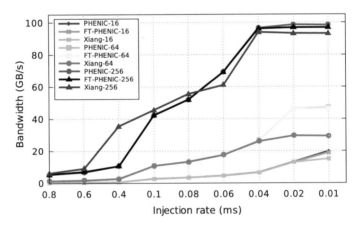

Fig. 5.37 Bandwidth comparison results under random uniform traffic

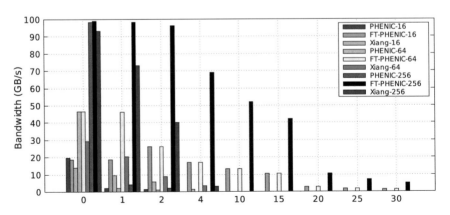

Fig. 5.38 Bandwidth comparison results as faults are introduced

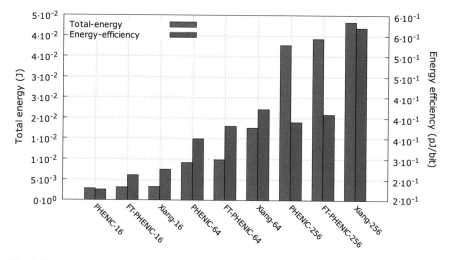

Fig. 5.39 Total energy and energy efficiency comparison results under random uniform traffic near saturation

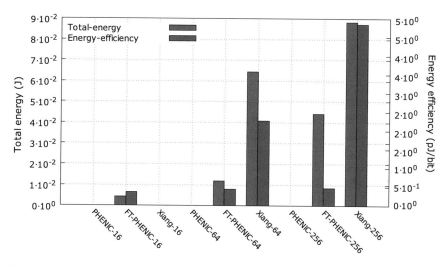

Fig. 5.40 Total energy and energy efficiency comparison results under random uniform traffic with 4% of MRs acting faulty

to notice how much the scale has changed for the energy efficiency between the fault-free and 4% fault results.

From these results, we can see that FT-PHENIC outperforms systems with either non-blocking or blocking switches. In addition, it provides heightened energy efficiency, far greater than the torus-based which can offer the same bandwidth as the proposed system. We conclude that the obtained improvement by FT-PHENIC is the result of the association of three main factors together: (1) the non-blocking switch supporting optical acknowledgment signals, (2) the light-weight router with reduced buffer size, (3) and the path setup algorithm to adopt hybrid switching inside the photonic switch.

5.3.6 Related Literature

There are three main types of optical fault tolerance that we were able to find. The first one is various methods of adaptive routing. The second one is techniques involving redundancy, which is commonly implemented in the network interface by using WDM as a redundancy technique. The third one involves buffering, checking, and proceeding like a standard electronic NoC.

Adaptive routing [94–96] is the most common method for fault tolerance in mesh-based architectures because of the large amount of possible minimal paths. It does require some extra logic in the routing decision, but this is minimal compared to an extra interconnect at each location. For it to truly support multiple faults, it must also support non-minimal routing in order to avoid a non-reserved deadlock situation. It should also be noted that implementing fault tolerance on a deadlock-free algorithm can negate that feature. This is not troublesome to optical networks as deadlock is a non-issue due to the fact that end-to-end is reserved before the transmission can start, and is only an issue during path setup.

Ramesh et al. proposed a method [95] of determining and using backup routes. The algorithm determines the least cost path. This path will be used unless there is a fault detected, in which case the backup path is used. Ramesh proposed using a set of probe packets. When the destination receives one of the probe packets, it then sends a PACK signal for each probe packet. If a packet is dropped due to faults, then a NACK signal is sent. This is a solution of off-chip optical networks though.

Loh breaks his algorithm [94] into a similar fashion to Ramesh. It has a default routing algorithm and a backup routing method. His two methods are called logical route and adaptive route. The logical route in his paper is a few sets of dimension-order routing. The adaptive algorithm determines which of the deterministic routings to use. This method simply checks for faults along the way, and if it can be detected, then it tries to switch to the other form of dimension-order routing. This is an attempt to shift from X to Y when a problem is found in the X-direction. This results in a routing algorithm which is minimal and adaptive, deadlock-free, and livelock-free.

Fault region [96] is a form of adaptive routing where each node keeps track of the permanent faults of its neighbors. This then allows for the path making decision to be educated with respect to faults up to a certain distance away. It can then guarantee that no old permanent faults are going to cause problems with the transmission. One such an algorithm is proposed by Xingyun [96]. He proposed a quite interesting optical network. It comes in the form of a torus which only allows data in two directions. This allows for some unique fault tolerance ideas. While they may not be minimalistic routing it will switch directions, go under the chip and come back from the top and reroute to avoid a bad crossing. This could possibly cause large amounts of insertion loss from routing around the network's length multiple times. This loss would translate to high power cost, and not yield any true benefits to converting to optical. This is still only monitoring its own outputs though.

Look-ahead routing [63] is another type of adaptive routing which is most interesting to implement in a nanophotonic setting. This is where a node has knowledge of its neighbors' faulty links, and possibly its neighbors' neighbors' links. With this data at hand, the routing can protect a path and guarantee its success. The only issue would be implementing one of the detection algorithms mentioned at the beginning of this section. Although it has not been implemented in a photonic chip yet, there is no obvious reason preventing it from being translated over. Xiang's method [63] uses a minus-first routing algorithm as a basis. The author does not detail how to detect a faulty link, but once a faulty link is discovered, it runs a minus-first algorithm, checking each step along the way. This method attempts to find all paths from the source to the destination from the problematic node, and then determines which one requires the least amount of time. This switch shows that only the links are optical, and the switches themselves are electrical. This also allows for the implementation of buffers, which allow for a few more fault tolerance options which can be detailed in Radetzki's paper [72].

Modular redundancy uses WDM (Wavelength Division Multiplexing) as a fault tolerance tool [97–99]. The general idea is that if a certain wavelength is causing problems, either through noise or a manufacturing defect, and this problem can be detected, then certain wavelengths can be disabled and enabled. This is highly effective for modulator and photodetector based faults. These focus on permanent and intermittent faults, because a transient fault would occur far too late for a wavelength to be switched.

Noise has been a large source of faults within optical networks. Currently, there are many different forms of optical switches which are used in networks. The main goal of these switch designs is to reduce the area, when compared to the crossbar switch. We will only focus on the non-blocking switches because of their performance benefits. Three examples of optical switches can be seen in Fig. 5.41. The first is an example of a typical optimized switch, which reduces crossings and MRs. The second is the five-port crossbar switch, which uses the maximum number of MRs, crossings, and terminators, but is a simplistic non-blocking design. The last, Crux by Ye et al. [81],

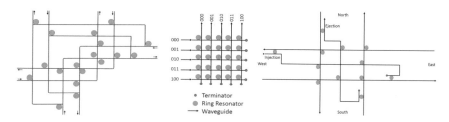

Fig. 5.41 Example of photonic switches. From left to right: PHENIC's original [48], crossbar, and crux [81]

is a switch which is optimized for XY-deterministic routing. This allows it to drop some extra MRs, but it no longer maintains the functionality to travel from the Y-direction to X-direction, such as north to east. This does greatly reduce the noise, when compared to other switches which can perform all network routing operations. Many other switches and networks were proposed to improve the SNR [78, 100–103]. The reason that this noise is so heavily researched is explained by Nikdast et al. [104].

Additionally, various authors have looked into the affect of *thermal variance*, and how to combat it [105, 106]. There are various ways to combat it, but the most common way is to cool down the ring to normal temperatures, which can be done by keeping it inactive, or by thermal tuning [106]. Trimming [60] was also one solution, which was mentioned in the introduction, and appears to be a promising answer to the problem. To the best of our knowledge, none of the existing solutions proposed so far take advantage of switch structure to provide fault tolerance. The focus of all other research has been on the routing algorithms or different locations to provide modular redundancy or noise reduction.

5.3.7 Chapter Summary

With the emergence of 3D integration technologies, a new opportunity emerges for chip architects by porting the 2D-NoC to the third dimension. In 3D integration technologies, multiple layers of active devices are stacked above each other and vertically interconnected using through-silicon via (TSV). As compared to 2D-IC designs, 3D-ICs allow for performance enhancements even in the absence of scaling because of the reduced interconnect lengths. In addition, package density is increased, power consumption is reduced, and the system is more immune to noise.

This chapter first presented architecture, design, and evaluation of a 3D-NoC including complete hardware design details about the main components of the 3D-NoC system. Then, the chapter presented a fault-tolerant photonic network-on-chip architecture, which uses minimal redundancy to assure accuracy of the packet transmission even after faulty microrings (MRs) are detected. The system is based on a fault-tolerant path-configuration and routing algorithm, and a microring fault-resilient photonic router.

References

1. A.B. Ahmed, High-performance scalable photonics on-chip network for many-core systems-on-chip, Ph.D. Thesis, Gradute school of Computer Science and Engineering, The University of Aizu, March 2016
2. A.B. Ahmed, A. Ben Abdallah, Graceful deadlock-free fault-tolerant routing algorithm for 3d network-on-chip architectures. J. Parallel Distrib. Comput. **74**(4), 2229–2240 (2014)

3. K. Kim, H.Y. Kim, T.G. Kim, Top-down retargetable framework with token-level design for accelerating simulation time of processor architecture. IEICE Trans. Fundam. Electron. Commun. Comput. Sci. **E86-A**(12), 3089–3098 (2003)
4. J. Kim, C. Nicopoulos, D. Park, V. Narayanan, M.S. Yousif, C.R. Das, A gracefully degrading and energy-efficient modular router architecture for on-chip networks, in *Proceedings of the 33rd International Symposium on Computer Architecture* (2006), pp. 138–149
5. R. Mullins, A. West, S. Moore, Low-latency virtual-channel routers for on-chip networks, in *Proceedings of the 31st International Symposium on Computer Architecture* (2004), pp. 188–197
6. W.J. Dally, Express cubes: improving the performance of kary-n-cube interconnection networks. IEEE Trans. Comput. **40**(9), 1016–1023 (1991)
7. J. Kim, J. Balfour, W.J. Dally, Flattenned butterfly topology for on-chip networks, in *Proceedings of the 40th International Symposium on Microarchitecture* (2007), pp. 172–182
8. U.Y. Orgas, R. Marculescu, It's a small world after all: NoC performance optimization via long-range link insertion. IEEE Trans. on VLSI Sys. **14**(7), 693–706 (2006)
9. G. Philip, B. Christopher, P. Ramm, Handbook of 3d Integration: Technology and Applications of 3d Integrated Circuits (Wiley-VCH, 2008)
10. S. Das et al., Technology, performance, and computer aided design of three-dimensional integrated circuits, in *Proceedings of the International Symposium on Physical Design* (2004)
11. P. Morrow, M. Kobrinsky, S. Ramanathan, C.-M. Park, M. Harmes, V. Ramachandrarao, H. Park, G. Kloster, S. List, S. Kim, Wafer-level 3d interconnects via cu bonding, in *Proceedings of the 21st Advanced Metallization Conference* (2004)
12. J. Joyner, P. Zarkesh-Ha, J. Meindl, A stochastic global net-length distribution for a three-dimensional system-on-chip(3D-SoC), in *Proceedings of the 14th Annual IEEE International ASIC/SOC Conference* (2001)
13. A.W. Topol, J.D.C. La Tulipe, L. Shi, D.J. Frank, K. Bernstein, S.E. Steen, A. Kumar, G.U. Singco, A.M. Young, K.W. Guarini, M. Ieong, Three-dimensional integrated circuits. IBM J. Res. Dev. **50**(4/5), 491–506 (2006)
14. L.P. Carloni, P. Pande, Y. Xie, Networks-on-chip in emerging interconnect paradigms: advantages and challenges, in *Proceedings of the 3rd ACM/IEEE International Symposium on Networks-on-Chip (NOCS09), San Diego, CA* (2009), pp. 93–102
15. F. Li, C. Nicopoulos, T.D. Richardson, Y. Xie, N. Vijaykrishnan, M.T. Kandemir, Design and management of 3d chip multiprocessors using network-in-memory, ISCA (2006), pp. 130–141
16. A. Ben Abdallah, M. Sowa, Basic network-on-chip interconnection for future gigascale mcsocs applications: communication and computation orthogonalization, in *Proceedings of Tunisia-Japan Symposium on Society, Science and Technology (TJASSST)* (2006), pp. 4–9
17. K. Mori, A. Ben Abdallah, K. Kuroda, Design and evaluation of a complexity effective network-on-chip architecture on FPGA, in *The 19th intelligent system symposium (FAN 2009)* (2009), pp. 318–321
18. K. Mori, A. Esch, A. Ben Abdallah, K., Kuroda, Advanced design issue for OASIS network-on-chip architecture, in *IEEE, International Conference on BWCCA* (2010), pp. 74–79
19. A.B. Ahmed, High-throughput architecture and routing algorithms towards the design of reliable mesh-based many-core network-on-chip systems, Ph.D. Thesis, Graduate School of Computer Science and Engineering, University of Aizu, March 2015
20. A. Habibi, M. Arjomand, H. Sarbazi-Azad, Multicast-aware mapping algorithm for on-chip networks, in *19th International Euromicro Conference on Parallel, Distributed and Network-Based Processing* (2011), pp. 455–462
21. G. Leary, K.S. Chatha, Design of NoC for SoC with multiple use cases requiring guaranteed performance, in *23rd International Conference on VLSI Design* (2010), pp. 200–205
22. R. Kumar, V. Zyuban, D.M. Tullsen, Interconnections in multicore architectures: understanding mechanisms, overheads and scaling, in *Proceedings of the 32nd International Symposium on Computer Architecture* (Madison, USA, 2005), pp. 408–419
23. B. Feero, P. Pratim Pande, Performance evaluation for three-dimensional networks-on-chip, in *Proceedings of IEEE Computer Society Annual Symposium on VLSI (ISVLSI)*, 9–11 May 2007, pp. 305–310

24. V.F. Pavlidis, E.G. Friedman, 3-D topologies for networks-on-chip. IEEE Trans. VLSI Syst. (2007), pp. 1081–1090
25. F. Li, C. Nicopoulos, T. Richardson, Y. Xie, V. Narayanan, M. Kandemir, Design and management of 3D chip multiprocessors using network-in-memory. ACM SIGARCH Comput. Archit. News **34**(2), 130–141 (2006)
26. S. Yan, B. Lin, Design of application-specific 3D networks-on-chip architectures, in *Proceedings of International Conference of Computer Design* (2008), pp. 142–149
27. C.J. Glass, L.M. Ni, The turn model for adaptive routing, in *Proceedings of the 19th Annual Intl Symposium on Computer Architecture* (1992), pp. 278–287
28. Z.S. Hu, F.Y. Hung, K.J. Chen, S.-J. Chang, W.-K. Hsieh, T.-Y. Liao, Improvement in thermal degradation of zno photodetector by embedding silver oxide nanoparticles. Funct. Mater. Lett. **6**(01), 1350001 (2013)
29. R.S. Ramanujam, B. Lin, Near-optimal oblivious routing on three dimensional mesh networks, in *Proceedings of the IEEE International Conference on Computer-Aided Design, Lake Tahoe, CA* (2008)
30. C.H. Chao, K.Y. Jheng, H.Y. Wang, J.C. Wu, A.-Y. Wu, Traffic- and thermal-aware run-time thermal management scheme for 3D NoC systems, in *Proceedings of ACM/IEEE International Symposium Networks-on-Chip (NoCS), Grenoble, France* (2010), pp. 223–230
31. S. Tyagi, Extended balanced dimension ordered routing algorithm for 3d-networks, in *Centre for Development of Advance Computing, Noida, (U.P.), India International Conference on Parallel Processing Workshops* (2009), pp. 499–506, http://www.iacqer.com/Proceedings
32. J.M. Montana, M. Koibuchi, H. Matsutani, H. Amano, Balanced dimension-order routing for k-ary n-cubes, in *Department of Information and Computer Science, Keio University, Yokohama, Japan, International Conference on Parallel processing Workshops* (2009), pp. 499–506
33. K. Dev, *Multi-objective Optimization Using Evolutionary Algorithms* (Wiley, New York, 2002), pp. 245–253
34. K. Lahiri, A. Raghunathan, S. Dey, Efficient exploration of the SoC communication architecture design space, in *Proceedings of IEEE/ACM ICCAD'00* (2000), pp. 424–430
35. L. Xin, C.S. Choy, Low-latency NoC Router with Lookahead Bypass, in *Proceedings of 2010 IEEE International Symposium on Circuits and Systems (ISCAS)* (2010), pp. 3981–3984
36. A. Ben Ahmed, A. Ben Abdallah, K. Kuroda, Architecture and design of efficient 3D network-on-chip (3D NoC) for custom multicore SoC, in *IEEE Proceedings of BWCCA-2010* (2010)
37. M.S. Rasmussen, Network-On-Chip in Digital Hearing Aids, Informatics and Mathematical Modeling, Technical University of Denmark, DTU, Richard Petersens Plads, Building 321, DK-2800 Kgs. Lyngby, IMM-Thesis-2006-76 (2006)
38. A. Pullini, F. Angiolini, D. Bertozzi, L. Benini, Fault tolerance overhead in network-on-chip flow control schemes, in *Proceedings of the Symposium on Integrated Circuits and Systems Design* (2005), pp. 224–229
39. Z. Fu, X. Ling, The design and implementation of arbiters for network-on-chips, in *2010 2nd International Conference on IEEE, Industrial and Information Systems (IIS)*, vol. 1 (2010) pp. 292–295
40. B.T. Gold, Balancing Performance, Area, and Power in an On-Chip Network., Master's thesis, Department of Electrical and Computer Engineering, Virginia Tech, Aug 2004
41. J. Rosethal, JPEG image compression using an FPGA, Master of Science in Electrical and Computer Engineering, University of California Santa Barbara DEC (2006)
42. S. Mandal, N. Gupta, A. Mandal, J. Malave, J. Lee, R. Mahapatra, NoCBench: a Benchmarking Platform for Network on Chip, in Workshop on Unique Chips and Systems (UCAS) (2009)
43. A.B. Ahmed, A. Ben Abdallah. Phenic: silicon photonic 3d-network-on-chip architecture for high-performance heterogeneous many-core system-on-chip, in *2013 14th International Conference on Sciences and Techniques of Automatic Control and Computer Engineering (STA)* (2013), pp. 1–9
44. A.B. Ahmed, M. Meyer, Y. Okuyama, A. Ben Abdallah, Efficient router architecture, design and performance exploration for many-core hybrid photonic network-on-chip (2d-phenic),

in *2015 2nd International Conference on Information Science and Control Engineering (ICISCE)* (2015), pp. 202–206

45. A.B. Ahmed, M. Meyer, Y. Okuyama, A. Ben Abdallah, Hybrid photonic noc based on non-blocking photonic switch and light-weight electronic router, in *2015 IEEE International Conference on Systems, Man and Cybernetics (SMC)* (2015)

46. A.B. Ahmed, Y. Okuyama, A. Ben Abdallah, Contention-free routing for hybrid photonic mesh-based network-on-chip systems, in T*he 9th IEEE International Symposium on Embedded Multicore/Manycore SoCs (MCSoc)* (2015), pp. 235–242

47. A.B. Ahmed, Y. Okuyama, A. Ben Abdallah, Non-blocking electro-optic network-on-chip router for high-throughput and low-power many-core systems, in *The World Congress on Information Technology and Computer Applications 2015* (2015)

48. A.B. Ahmed, A. Ben Abdallah, Hybrid silicon-photonic network-on-chip for future generations of high-performance many-core systems. J. Supercomput. (2015). doi:10.1007/s11227-015-1539-0

49. S. Zhu, G.-Q. Lo, Vertically-stacked multilayer photonics on bulk silicon toward three-dimensional integration. J. Lightw. Technol. **PP**(99), 1–1 (2015)

50. M.C. Meyer, A.B. Ahmed, Y. Okuyama, A. Ben Abdallah, Fttdor: microring fault-resilient optical router for reliable optical network-on-chip systems, in *2015 IEEE 9th International Symposium on Embedded Multicore/Many-core Systems-on-Chip (MCSoC)* (2015), pp. 227—234

51. B.R. Koch, A.W. Fang, O. Cohen, J.E. Bowers, Mode-locked silicon evanescent lasers. Opt. Express **18**(15), 11225 (2007)

52. A. Kumar, L.-S. Peh, P. Kundu, N.K. Jha, Express virtual channels: towards the ideal interconnection fabric, in *Proceedings of the 34th International Symposium on Computer Architecture* (2007), pp. 150–161

53. V.R. Almeida et al., All-optical switching on a silicon chip. Opt. Lett. **29**, 2867–2869 (2004)

54. S. Parsricha, N. Dutt, Trends in emerging on-chip interconnect technologies. *IPSJ Transaction on System LSI Design Methodology* vol. 1, pp. 2–17 (2008)

55. M. Briere, et. al., Heterogeneous modelling of an optical network-on-chip with SystemC, in *The 16th IEEE International Workshop on Rapid System Prototyping* 8–10 June 2005, pp. 10–16

56. J. Chan et al., PhoenixSim: A Simulator for Physical-Layer Analysis of Chip-scale Photonic Interconnection Networks, in Design, Automation and Test in Europe (DATE) (2010)

57. R. Kappeler, Radiation testing of micro photonic components. Stagiaire Project Report. ESA/ESTEC. Sept.29, 2004. Ref. No.: EWP 2263

58. W. Bogaerts, P. De Heyn, T. Van Vaerenbergh, K. De Vos, S. Kumar Selvaraja, T. Claes, P. Dumon, P. Bienstman, D. Van Thourhout, R. Baets, Silicon microring resonators. Laser Photonics Rev. **6**(1), 47–73 (2012)

59. C.J. Nitta, M.K. Farrens, V. Akella, Resilient microring resonator based photonic networks, in *Proceedings of the 44th Annual IEEE/ACM International Symposium on Microarchitecture, MICRO-44* (ACM, New York, 2011), pp. 95–104

60. J.H. Ahn, M. Fiorentino, R.G. Beausoleil, N. Binkert, A. Davis, D. Fattal, N.P. Jouppi, M. McLaren, C.M. Santori, R.S. Schreiber, S.M. Spillane, D. Vantrease, Q. Xu, Devices and architectures for photonic chip-scale integration. Appl. Phys. A **95**(4), 989–997 (2008)

61. S.T. Chu, W. Pan, S. Sato, T. Kaneko, B.E. Little, Y. Kokubun, wavelength trimming of a microring resonator filter by means of a uv sensitive polymer overlay. Photonics Technol. Lett. IEEE **11**(6), 688–690 (1999)

62. D. Rafizadeh, J.P. Zhang, S.C. Hagness, A. Taflove, K.A. Stair, S.T. Ho, R.C. Tiberio, Temperature tuning of microcavity ring and disk resonators at 1.5- mu;m, in *Lasers and Electro-Optics Society Annual Meeting, 1997. LEOS '97 10th Annual Meeting. Conference Proceedings, IEEE*, vol. 2 (1997), pp. 162–163

63. D. Xiang, Y. Zhang, S. Shan, Y. Xu, A fault-tolerant routing algorithm design for on-chip optical networks, in *2013 IEEE 32nd International Symposium Reliable Distributed Systems (SRDS)* (2013), pp. 1–9

64. Y. Xu, J. Yang, R. Melhem, Tolerating process variations in nanophotonic on-chip networks, in *ACM SIGARCH Computer Architecture News*, vol. 40 (IEEE Computer Society, 2012), pp. 142–152
65. Z. Tu, Z. Zhou, X. Wang. Reliability considerations of high speed germanium waveguide photodetectors, in *SPIE OPTO* (International Society for Optics and Photonics, 2014), pp. 89820W–89820W
66. S.g. Yang, L. Li, Y. a. Zhang, B. Zhang, Y. Xu, A power-aware adaptive routing scheme for network on a chip, in *7th International Conference on ASIC, 2007. ASICON '07* (2007), pp. 1301–1304
67. J. Keane, C.H. Kim, An odometer for cpus: microprocessors don't normally show wear and tear, but wear they do. IEEE SPECTRUM **48**(5), 26–31 (2011)
68. S. Luryi, J. Xu, A. Zaslavsky, *Future Trends in Microelectronics: Up the Nano Creek* (Wiley, 2007)
69. M. Agarwal, B.C. Paul, M. Zhang, S. Mitra, Circuit failure prediction and its application to transistor aging, in *25th IEEE VLSI Test Symposium (VTS'07), IEEE* (2007), pp. 277–286
70. J. Keane, T.-H. Kim, C. H Kim, An on-chip nbti sensor for measuring pmos threshold voltage degradation. IEEE Trans. Very Large Scale Integr. (VLSI) Sys. **18**(6), 947–956 (2010)
71. E. Mintarno, J. Skaf, R. Zheng, J.B. Velamala, Y. Cao, S. Boyd, R.W Dutton, S. Mitra, Self-tuning for maximized lifetime energy-efficiency in the presence of circuit aging. IEEE Trans. Comput.-Aided Des. Integr. Circuits Syst. **30**(5), 760–773 (2011)
72. M. Radetzki, C. Feng, X. Zhao, A. Jantsch, Methods for fault tolerance in networks-on-chip. ACM Comput. Surv. (CSUR) **46**(1), 8 (2013)
73. K. Kuhn, C. Kenyon, A. Kornfeld, M. Liu, A. Maheshwari, W.-k. Shih, S. Sivakumar, G. Taylor, P. VanDerVoorn, K. Zawadzki, Managing process variation in intel's 45nm cmos technology. Intel Tech. J. **12**(2) (2008)
74. S.K. Saha, Modeling process variability in scaled cmos technology. IEEE Des. Test Comput. **27**(2), 8–16 (2010)
75. M. Meyer, Micro-ring fault-resilient photonic on-chip network for reliable high-performance many-core systems-on-chip, Ph.D. Thesis, Graduate School of Computer Science and Engineering, The University of Aizu, March 2017
76. A.B. Ahmed, A. Ben Abdallah, Architecture and design of high-throughput, low-latency, and fault-tolerant routing algorithm for 3d-network-on-chip (3d-noc). J. Supercomput. **66**(3), 1507–1532 (2013)
77. M. Nikdast, G. Nicolescu, S. Le Beux, J. Xu, *Photonic Interconnects for Computing Systems* River Publishers Series (2017). ISBN:9788793519800
78. M. Mohamed, Silicon Nanophotonics for Many-Core On-Chip Networks. Ph.D. thesis, University of Colorado (2013)
79. K. Preston, N. Sherwood-Droz, J.S. Levy, M. Lipson, Performance guidelines for wdm interconnects based on silicon microring resonators, in *2011 Conference on Lasers and Electro-Optics (CLEO)* (2011), pp. 1–2
80. L. Brusberg, H. Schrder, M. Queisser, K.-D. Lang, Single-mode glass waveguide platform for dwdm chip-to-chip interconnects, in *Electronic Components and Technology Conference (ECTC), 2012 IEEE, 62nd* (2012), pp. 1532–1539
81. Y. Ye, X. Wu, J. Xu, W. Zhang, M. Nikdast, X. Wang, Holistic comparison of optical routers for chip multiprocessors, in *2012 International Conference on Anti-Counterfeiting, Security and Identification (ASID)* (IEEE, 2012), pp. 1–5
82. J. Chan, G. Hendry, A. Biberman, K. Bergman, L.P Carloni, Phoenixsim: a simulator for physical-layer analysis of chip-scale photonic interconnection networks, in *Proceedings of the Conference on Design, Automation and Test in Europe* (European Design and Automation Association, 2010), pp. 691–696
83. A. Ben Abdallah, *Multicore Systems-On-chip: Practical Hardware/Software Design*, 2nd edn. (Atlantis, Paris, 2013)
84. A. Ben Abdallah, M. Sowa, Basic network-on-chip interconnection for future gigascale MCSoCs applications: communication and computation orthogonalization, in *JASSST2006* (2006)

85. C.A.D. Adi, H. Matsutani, M. Koibuchi, H. Irie, T. Miyoshi, T. Yoshinaga, An efficient path setup for a photonic network-on-chip, in *2010 First International Conference on Networking and Computing* (2010), pp. 156–161

86. J. Chan, G. Hendry, K. Bergman, L.P. Carloni, Physical-layer modeling and system-level design of chip-scale photonic interconnection networks. IEEE Trans. Comput.-Aided Design Integr. Circuits Syst. **30**(10), 1507–1520 (2011)

87. G. Hendry, E. Robinson, V. Gleyzer, J. Chan, L.P. Carloni, N. Bliss, K. Bergman, Circuit-switched memory access in photonic interconnection networks for high-performance embedded computing, in *2010 International Conference on High Performance Computing, Networking, Storage and Analysis (SC)* (2010), pp. 1–12

88. A. Shacham, K. Bergman, L.P. Carloni. On the design of a photonic network-on-chip, in *Networks-on-Chip, 2007. NOCS 2007. First International Symposium* (2007), pp. 53–64

89. A.B. Kahng, B. Li, L.-S. Peh, K. Samadi, Orion 2.0: a power-area simulator for interconnection networks. IEEE Trans. Very Large Scale Integr. (VLSI) Sys. **20**(1), 191–196 (2012)

90. J. Chan, K. Bergman, Photonic interconnection network architectures using wavelength-selective spatial routing for chip-scale communications. IEEE/OSA J. Opt. Commun. Netw. **4**(3), 189 (2012)

91. Y. Pan, P. Kumar, J. Kim, G. Memik, Y. Zhang, A. Choudhary, Firefly: illuminating future network-on-chip with nanophotonics, in *ACM SIGARCH Comput. Archit. News*, vol. 37 (ACM, 2009), pp. 429–440

92. A. Shacham, K. Bergman, L.P. Carloni, Photonic networks-on-chip for future generations of chip multiprocessors. IEEE Trans. Comput. **57**(9), 1246–1260 (2008)

93. D. Vantrease, R. Schreiber, M. Monchiero, M. McLaren, N.P Jouppi, M. Fiorentino, A. Davis, N. Binkert, R. G Beausoleil, J.H. Ahn, Corona: system implications of emerging nanophotonic technology, in *ACM SIGARCH Computer Architecture News*, vol. 36 (IEEE Computer Society, 2008), pp. 153–164

94. P.K.K. Loh, W.-J. Hsu, Design of a viable fault-tolerant routing strategy for optical-based grids, in *Parallel and Distributed Processing and Applications* (Springer, 2003), pp. 112–126

95. G. Ramesh, S. SundaraVadivelu, A reliable and fault tolerant routing for optical wdm networks (2009), arXiv:0912.0602

96. Q. Xingyun, F. Quanyou, C. Yongran, D. Qiang, D. Wenhua, A fault tolerant bufferless optical interconnection network, in *Eighth IEEE/ACIS International Conference on Computer and Information Science, 2009. ICIS 2009* (IEEE, 2009), pp. 249–254

97. M. McLaren, N.L. Binkert, A.L.Davis, M.Florentino, Energy-efficient and fault-tolerant resonator-based modulation and wavelength division multiplexing systems, 22 2014. US Patent 8,705,972

98. L. Sahasrabuddhe, S. Ramamurthy, B. Mukherjee, Fault management in ip-over-wdm networks: Wdm protection versus ip restoration. IEEE J. Sel. Area. Comm. **20**(1), 21–33 (2002)

99. J. Zhang, B. Mukheriee, A review of fault management in wdm mesh networks: basic concepts and research challenges. IEEE Netw. **18**(2), 41–48 (2004)

100. S.V.R. Chittamuru, S. Pasricha, Crosstalk mitigation for high-radix and low-diameter photonic noc architectures. Design Test, IEEE **32**(3), 29–39 (2015)

101. S.V.R. Chittamuru, S. Pasricha, Improving crosstalk resilience with wavelength spacing in photonic crossbar-based network-on-chip architectures, in *2015 IEEE 58th International Midwest Symposium on Circuits and Systems (MWSCAS)* (IEEE, 2015), pp. 1–4

102. P.K. Kaliraj, Reliability-performance trade-offs in photonic noc architectures (2013)

103. M. Nikdast, X. Jiang, W. Xiaowen, W. Zhang, Y. Ye, X. Wang, Z. Wang, Z.Wang, Systematic analysis of crosstalk noise in folded-torus-based optical networks-on-chip. IEEE Trans. Comput.-Aided Design Integr. Circuits Syst. **33**(3), 437–450 (2014)

104. M. Nikdast, J. Xu. On the impact of crosstalk noise in optical networks-on-chip, in *Design Automation Conference (DAC)* (2014)

105. H. Li, A. Fourmigue, S.L. Beux, X. Letartre, I. O'Connor, G. Nicolescu, Thermal aware design method for vcsel-based on-chip optical interconnect, in *Proceedings of the 2015 Design, Automation and Test in Europe Conference and Exhibition* (EDA Consortium, 2015), pp. 1120–1125

106. Z. Li, M. Mohamed, X. Chen, E. Dudley, K. Meng, L. Shang, A.R. Mickelson, R. Joseph, M. Vachharajani, B. Schwartz et al., Reliability modeling and management of nanophotonic on-chip networks. IEEE Trans. Very Large Scale Integr. (VLSI) Syst. **20**(1), 98–111 (2012)

Chapter 6
3D Integration Technology for Multicore Systems On-Chip

Abstract 3D integration fully explains the latest microelectronics techniques for increasing chip density and maximizing performance while reducing power consumption. Three-dimensional NoCs/SoCs systems have been showing their advantages against conventional two-dimensional SoCs. Thanks to their reduced average interconnect length and lower interconnect-power consumption inherited from three-dimensional ICs. To ensure their correct functionality, such systems must be fault-tolerant to any short-term malfunction or permanent physical damage to ensure message delivery on time while minimizing the performance degradation as much as possible. This chapter introduces 3D integration technology for fault-tolerant multicore Systems On-Chip.

6.1 3D Integration Technology

During the past few decades, a lot of research has been focusing on three-dimensional multicore SoCs/NoCs as an auspicious solution to alleviate the interconnect bottleneck and reduce the power consumption in current SoCs designs. By increasing system integration at a lower cost, reducing footprint, improving the performance and reusing the existed technologies, 3D integration is a promising approach for manufacturing future advanced ICs. Recently, 3D-ICs have been introduced in several applications such as DRAM stacking [1], camera sensors [2–4], SSD (Solid State Drive) [5], processor [6].

By stacking multiple 2D layers together, designers expect to have smaller package, shorter wire, and better overall performance. Figure 6.1 illustrates how 3D integration can reduce the footprint and wire length. We can notice that with the same area A, the die width of 2D-IC is $\sqrt{A}$ while the two-layer and four-layer 3D-IC dies width are $\sqrt{A/2}$ and $\sqrt{A/4}$ (29.29 and 50% of reduction), respectively. Thanks to the reduction of the die width, the wire length is also reduced.

As shown in Fig. 6.2, there are several methods for 3D integration, such as *wire bonding* [7], *solder balls* [8], *through-silicon-via* [9] and *wireless stacking* [10, 11]. The *Wire bonding* uses dedicated wires to connect cores within different layers. It has different design configurations—from pyramid stacking to overhang

© Springer Nature Singapore Pte Ltd. 2017
A. Ben Abdallah, *Advanced Multicore Systems-On-Chip*,
DOI 10.1007/978-981-10-6092-2_6

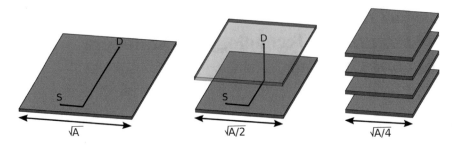

Fig. 6.1 Reducing footprint and wire length in 3D-stack structure

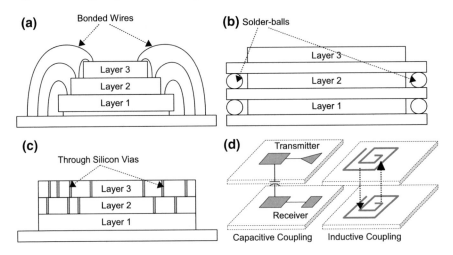

Fig. 6.2 3D integration schemes: **a** Wire bonding; **b** Solder balls; **c** Through silicon vias; **d** Wireless stacking

stacking, from standard bonding to low loop and reverse bonding. The *Solder balls* is an alternative method for wire bonding, and used solder balls to connect pins in different stacked layers. For both *wire bonding* and *solder balls* methods, the major problem is the global wire interconnections. The *through-silicon via (TSV)* is a vertical connection passing completely through a silicon wafer. TSVs are a high performance interconnect techniques used as an alternative to the previous described methods to create 3D packages and 3D integrated circuits. Finally, the *Wireless stacking* approach eliminates difficult steps needed for the TSV integration method, such as extreme thinning of wafers, however it opens up new issues about aligning coils and eliminating interference.

6.2 Fault-Tolerant TSV Cluster for 3D Integration

Numerous works have addressed the fault tolerance and reliability issues in 3D-NoCs/SoCs [12–15]. In this chapter, we focus on TSV defect tolerance. The existing works have approached the TSV fault-tolerance in three layers: *Physical layer*, *Data-link layer* and *System layer*.

In *Physical layer*, the improvement of TSV manufacturing can help to reduce the defect rate [16]. Designers can optimize the physical layout, use thermal-aware routing and placement methods to improve the reliability of 3D-ICs [17]. Even when a fabricated TSV has a short defect, a correction circuit, using a voltage comparator to gain the output voltage of the TSV, can be employed [18]. To enhance the reliability of TSVs, [19] proposed a method named *Double TSV* which uses two TSVs, instead of one, to maintain the vertical communication. If an open, short-to-substrate or bridge defect occurs in one TSV, the communication is still performed by the duplicate one.

In the *Data-link layer*, the most common method is adding redundant TSVs to correct the defected ones [20–22]. The major concern in this method is to efficiently route from a defected TSV to a spare one. There are four basic solutions: (a) signal switching [1], (b) single shifting [23], (c) crossbar [21] and (d) network routing [20]. Because of the cluster defect, the TSVs located nearby the defected one have a high probability of failure. Therefore, grouping them together will make a group with many defected TSVs. This leads to increase the redundancy to deal with this kind of defects. In [24], the authors propose a mapping method to reduce the impact of cluster defects. TSVs in the same group are mapped in a random position with the help of an optimization process. On the other hand, Zhao et al. [25] analyze the grouping method to achieve the best recovery. The work presented in [20] introduces an innovated method for TSV mapping by creating a network and implementing an algorithm for rerouting the defected TSVs.

Loi et al. [21], proposed a TSV fault-tolerant structure for 3D-NoCs. The authors propose the crossbar redundancy to enhance the yield rate. A testing mechanism is also presented to help the system detecting the defected TSVs. Because TSVs manage the vertical connections in a 3D-NoC, *Error Correction Coding* [26] is also a prominent method for detecting and correcting the defected TSVs; however, this type of solutions requires extra bits, which significantly increases the area cost and power consumption when compared to the redundancy approach.

In the *System layer*, which mainly focuses on 3D-NoCs, fault-tolerant routing algorithms [27] are one of the most suitable solutions. To reduce the risk of thermal and tress issues in 3D-NoCs, thermal-aware management [28] is also a promising solution. On the other hand, most of works proposed off-line testing and recovery schemes which are not suitable for post-manufacturing. The system operation has to be halt in order to be tested and recovered. In [22], the authors presented an online testing function. Because the reliability of TSVs is a critical issue, the need for online testing recovery is primordial.

As we previously mentioned, the cluster defect is predicted to be frequently occurred. The most efficient solution for correcting random defects is grouping and

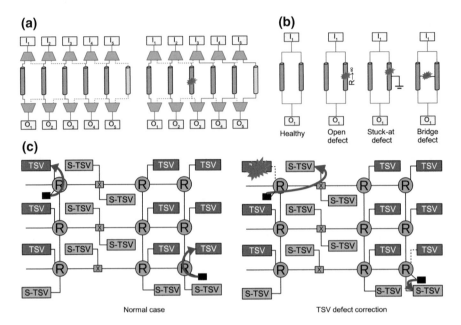

Fig. 6.3 TSV fault-tolerance schemes: **a** Redundancy technique; **b** Double TSV; **c** Network TSV

adding redundancy. However, they are still inefficient for the cluster defect and require costly extra area for redundancy. On the other hand, several works [29, 30] have been reporting the low utilization of the vertical connection using TSVs in 3D-NoC. The authors tried to reduce the number of TSVs to minimize the area overhead while maintaining a low degradation in terms of performance. Motivated by the cluster defect issue and the low utilization of the TSVs in 3D-NoC, we present in this chapter a low cost method for TSV fault-tolerance in 3D-NoCs. Figure 6.3 illustrates well-known TSV fault-tolerance schemes.

6.2.1   Fault-Tolerance for TSV-Clusters

In order to handle the TSV-cluster defects in 3D-NoCs, our solution is to share TSVs between neighboring routers. Therefore, when a TSV cluster fails, its router can borrow a healthy cluster from one of its neighbors to maintain the connection. Moreover, we also present several design optimizations to improve the reliability of the system (Sect. 6.3.4).

6.2.1.1   Fault Assumptions

Before we present the system structure, this subsection clarifies the fault assumptions taken in this proposal. Because the cluster defect [20, 24, 25] is the major obstacle to be dealt with in this work, we assume there are no random defects. Here, we consider an occurred fault makes the whole TSVs in the cluster defected. For those who might be concerned about random defects, using redundancy [1, 19, 21, 23] can be easily integrated in our TSV-cluster design. For controlling signals using TSVs, they are considered as a part of the TSV cluster instead of separated TSVs, which are better dealt as random defect (e.g., [22] uses Double TSV [19]). The detection process, which may need a Built-In-Self-Test module [31, 32], is assumed to be existing and connected to the fault-tolerance module. To synchronize the configuration, the existing NoC infrastructure is used instead of adding TSVs. Therefore, no redundancy is required in the proposed architecture.

6.2.1.2   System Structure

A simplified layout example of $3 \times 3 \times 3$ 3D-NoC system using the proposed TSV usage is depicted in Fig. 6.4. For each vertical connection, a router needs a set of TSVs. Instead of grouping all TSVs together they are divided into four groups. As a result, a router owns four TSV clusters and has a maximum of four nearby TSV clusters. If a TSV cluster of a router is defected, the router can choose one of its four neighboring clusters as a replacement without the need for redundancy. To satisfy the timing constraints, the router chooses the closest TSV cluster among its neighbor clusters. Taking into account further TSV clusters is not considered in order to avoid long wires that are needed to establish the connection. By structuring the TSVs into four clusters for each router, we can maintain the scalability of 3D-NoCs and avoid long wire delay. We have to note here that there are some works that consider serialization to reduce the cost of TSVs in 3D-NoCs. In this work, we consider a normal vertical connection; however, the proposed approach can be applied for the serialized TSV structure.

Figure 6.5 shows the placement and connections of the TSV sharing area between *R(1,1,1)* and *R(1,0,1)*. Because each router has two ports (up and down) and two directions (in and out), the number of TSV clusters is eight. Each TSV cluster handles a quarter of the vertical connection. By using the tri-stage gates, the system can control which router has access to the TSV clusters.

6.2.1.3   Sharing Circuit Design

To borrow a TSV cluster from a neighbor, the router needs a supporting module. Figure 6.6a shows the wrapper of a 3D-Router with the additional supporting modules that perform the sharing algorithm, later explained in Sect. 6.3. There are two identical sharing modules (S-UP and S-DOWN) for the two vertical up and down connections

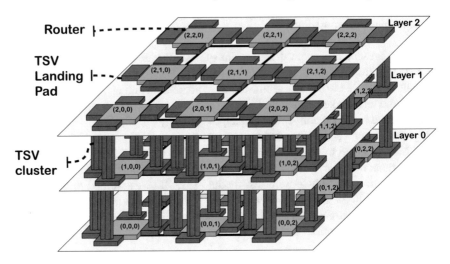

Fig. 6.4 High-level view of the system architecture with $3 \times 3 \times 3$ configuration

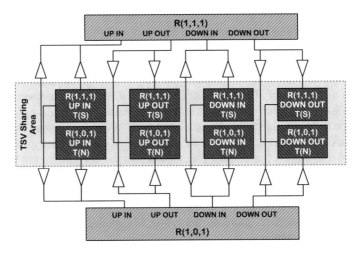

Fig. 6.5 TSV sharing area placement and connectivity between two neighboring routers

and each connection has two configuration registers (CR) for the input and output ports. As previously depicted in Fig. 6.4, $R(1,1,1)$ shares the TSV clusters with its four neighbors: $R(1,1,0)$, $R(1,1,2)$, $R(1,0,1)$, and $R(1,2,1)$. Figure 6.6b shows the sharing circuit for a TSV cluster. The input of this TSV cluster is shared between $R(2,1,0)$ and $R(2,1,1)$ on layer2. The output of this TSV cluster is shared between $R(1,1,1)$ and $R(1,1,0)$ on layer1. In the case where this TSV cluster is defected, or borrowed, the data can be sent by using one of the four neighboring clusters.

Based on the value of the 6-bit CR, shown in Table 6.1, the input and output ports can select the data from: (1) its original TSV cluster, (2) one of its four neighbor-

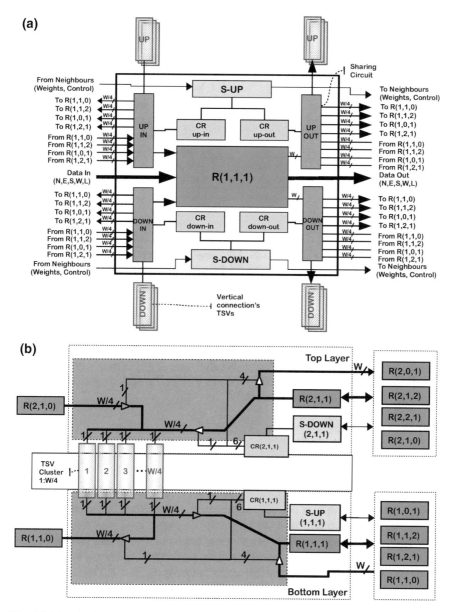

Fig. 6.6 The TSV fault-tolerance architecture: **a** Router wrapper; **b** Connection between two layers. *Red rectangles* represent TSVs. *S-UP* and *S-DOWN* are the sharing arbitrators which manage the proposed mechanism. *CR* stands for configuration register and W is the flit width

Table 6.1 Configuration register (CR) description

Value	Description
000001	Original router connects to the cluster
000010	Neighboring router connects to the cluster
000100	Original router connects to the neighboring north TSV cluster
001000	Original router connects to the neighboring east TSV cluster
010000	Original router connects to the neighboring south TSV cluster
100000	Original router connects to the neighboring west TSV cluster

ing clusters or (3) being disconnected. As shown in Fig. 6.6b, the output data from $R(2,1,1)$ can be sent to its TSV cluster if the least significant bit is "1". By setting the least significant bit to "0", the original TSV cluster is disconnected from it router. If the second bit is set as "1", the neighboring router $(R(2,1,0))$ takes the access to this cluster. When the original TSV cluster is defected or taken, the router needs to take one of its neighbor's clusters to maintain the connection based on the last 4-bit of CR. At the receiving router $(R(1,1,1))$, a similar CR is used to establish the connection. The value of this CR is identical to the sending router's CR. Because the CR only manages the connectivity, its value have to be set carefully to avoid the possible conflict of TSV-cluster usage and to optimize the performance. To this aim, an adaptive sharing algorithm is needed.

6.3   Adaptive Online TSV Sharing Algorithm

In the previous section, we presented how a router can use its nearby TSV clusters to maintain the connection and the operation on a layer. The CR values need to be configured in order to deal with the TSV defects. The simplest way for this process is to perform it offline and the configuration fuses the TSV group [20]. However, fixing the connections has two main drawbacks: (1) recovering a newly defected TSV needs to halt the system and perform again the mapping, and (2) each application has a different distribution in the vertical connections and variations depending on the running task which is not optimized by offline mappings. Consequently, we aim to perform the mapping online so that the system can react immediately to the newly defected TSV clusters and can consider the connectivity of the 3D-NoC system. Thus, this subsection provides an online algorithm for sharing TSVs which can be implemented onto the system.

Figure 6.7 shows the proposed algorithm for our sharing mechanism. Each router is assigned to a weight for each of the vertical connections. This weight decides its priority in sharing/borrowing. The weight can be assigned at the design process or

```
                // Weight values of the current router and its N neighbors
                Input: Weight_current, Weight_neighbor[1 : N]
                // Status of current and neighboring TSV-clusters
                Input: TSV_Status_current[1 : N], TSV_Status_neighbor[1 : N]
                // Request to link TSV-clusters to neighbors
                Output: RQ_link[1 : N]
                // Current router status
                Output: Router_Status

1   foreach TSV_Status_current[i] do
2       if TSV_Status_current[i] == "NORMAL" then
              // It is a healthy TSV-cluster
3             RQ_link[i] = "NULL"
4       else
              // It is a faulty or borrowed TSV-cluster
5             find c in 1:N with:
6             Weight_neighbor[c] < Weight_current
7             Weight_neighbor[c] is minimal
8             and TSV_Status_neighbor[c] == "NORMAL";
9             if (c==NULL) then
10                return RQ_link[i] = "NULL"
11                return Router_Status = "DISABLE"
12            else
13                return RQ_link[i] = c
14                return Router_Status = "NORMAL"
```

Fig. 6.7 Adaptive online TSV sharing algorithm

can be updated by a dedicated module. Changing the weights of routers can create different mappings. At the initial stage, all routers in the network exchange their weights and their TSV-clusters status with their neighbors. In the next step, the algorithm performs the mapping process. If a TSV cluster is defected, its corresponding router should find from its neighbors a possible candidate by relying on the following conditions:

- The weight of the candidate has to be smaller than the current router.
- The candidate TSV cluster has to be healthy and not borrowed.
- The weight of the final candidate is the smallest among all the possible candidates.

At the end of the algorithm, the router finds out the possible candidate for borrowing. If no candidates were found, the router's vertical connection is disabled. If there is a candidate, the router sends a request to the borrowing router to use its TSV cluster as a replacement for the defected one. The routers having borrowed TSV clusters also look for a replacement among one of their neighbors. By using a weighted system, the disabled TSV-clusters focus on smaller weight routers.

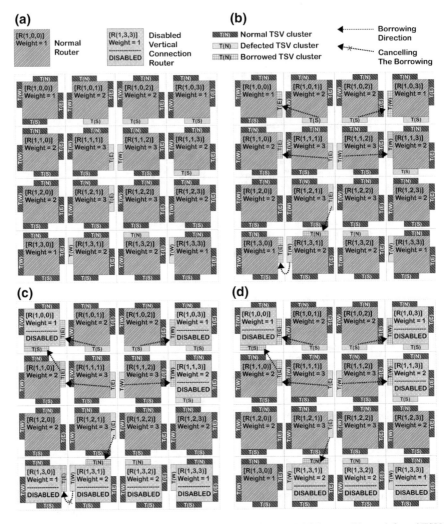

Fig. 6.8 An example of the sharing algorithm on a 4 × 4 layer: **a** Initial state with ten defected TSV clusters; **b** Best candidates selection; **c** Borrowing chain creating and selection refining. **d** Final result with six disabled routers

Figure 6.8 shows an example of how the sharing algorithm works on a 4 × 4 layer with ten defected TSV clusters. Initially, the routers in the center, which are predefined to have higher TSV utilization rates, have higher weights than those at the edges of the network, as depicted in Fig. 6.8a. The sharing algorithm selects the best candidates, shown in Fig. 6.8b, by following the rules previously explained in Fig. 6.7. Figure 6.8c shows that this selection must be further refined by disabling the router having less than four functional (or not borrowed) TSV clusters and canceling their borrowing. The returning process is discussed in Sect. 6.3.2. Moreover, we also observe the case in Fig. 6.8d where two routers $R(1,3,2)$ and $R(1,3,3)$ are disabled but $R(1,3,3)$ can borrow TSV cluster from $R(1,3,2)$ to obtain full four TSV clusters. However, the borrowing is prohibited due to the higher weight of router $R(1,3,2)$. In order to optimize this case, we use a technique named *Weight adjustment* in Sect. 6.3.3.

As shown in the above example, the chain of sharing leads to disabling the routers on the edges. Instead of having ten defected TSV clusters, the algorithm only disables six routers having the lowest weights (40% of reduction). Consequently, maintaining the connections of the center routers, which have higher weights and utilize more vertical communications, can reduce the impact of TSV defects in terms of overall performance.

6.3.1 Weight Generation

One of the most important parameters in the sharing algorithm is the weight values of the routers. The weights help the algorithm decide what router is suitable to be borrowed. As shown in Fig. 6.8, the routers having smaller weights are disabled after the chains of sharing are established.

Because the weights decide the priority of the routers in the sharing process, they need to be optimized to obtain a maximum system performance. In order to do that, the best solution is using a statistic-based solution where the priority of the vertical connection depends on the communication traffic [33, 34]. In other words, the vertical connections having more data transmissions are assigned higher weights; otherwise, smaller weights are assigned. Because application mapping is out of the scope of this work, we adopt a simple method where the routers in the middle of the layer have the highest weights. The router's weights are decreased and become the lowest at the edges of the layer. Equation 6.1 shows the used weight value assignment. The output of this weight assignment on a layer of 4×4 can be seen in Fig. 6.8 where, for instance, the weights of routers $R(1,0,0)$, $R(1,1,0)$, and $R(1,1,1)$ are 1, 2, and 3, respectively.

$$\text{Weight}_{\text{router}}(x, y) = \min(x, \text{cols} - x) + \min(y, \text{rows} - y) + 1 \qquad (6.1)$$

6.3.2 TSV-Clusters Return

After a TSV cluster is borrowed, it is managed by the borrowing router. However, if the borrowing router is disabled later, this frees the borrowed cluster which has to be returned to its original router. As a result, if the borrowed TSV cluster created a chain of borrowing, a chain of returning is also created. This can be clearly seen in Fig. 6.8c where $R(1,3,1)$ has a faulty cluster and has selected the east cluster of $R(1,3,0)$ to be borrowed. However, in the next step, $R(1,3,1)$ is selected to borrow its north cluster to a higher weight router, *R(1,2,1)*. Because *R(1,3,1)* is unable to find any sharing TSV cluster to borrow, it is disabled and borrowing from *R(1,3,0)* is canceled. Figure 6.8d represents the final results of the sharing process. In this final stage, *R(1,3,0)* is operational again as it is no longer lending a cluster to *R(1,3,1)* which was disabled in the previous phase.

After a TSV cluster is returned, its router check whether it created a borrowing chain and release the borrowing. If there is no borrowing chain, which means the router failed to find a replacement and is disabled, the sharing algorithm is performed again to check if the router can return to normal. As shown in Fig. 6.8d, *R(1,3,0)* returns to normal after its TSV cluster (T(E)) is returned.

6.3.3 Weight Adjustment

After applying the sharing mechanism, the disabled TSV clusters are shifted to the region which consists of low weighted routers. Figure 6.9a shows a case of three routers (*R(1,0,0)*, *R(1,0,1)* and *R(1,0,2)*) which are disabled after the sharing process. However, there are still a chance of optimizing these routers to obtain a better mapping. In fact, *R(1,0,2)* can borrow a TSV cluster from *R(1,0,1)*. Therefore, the number of TSV clusters of *R(1,0,2)* can be maintained to four.

To perform this optimization, the disabled router, after the sharing process by the algorithm shown in Fig. 6.7, is brought to a new process. First, the router counts the number of possible TSV clusters that it can borrow. Since three routers (*R(1,0,0)*, *R(1,0,1)* and *R(1,0,2)*) are disabled, their TSV clusters are free to be taken. At the end of this stage, *R(1,0,0)*, *R(1,0,1)* and *R(1,0,2)* have 1, 3, and 1 borrowed/defected TSV clusters and are able to take 0, 1 and 1 TSV cluster from their disabled neighbors, respectively. At the second stage, the router checks whether it can take the disabled router's cluster to obtain a full connection. Because *R(1,0,2)* has one borrowed cluster and is able to borrow another one from *R(1,0,1)*, its weight is kept. The other routers

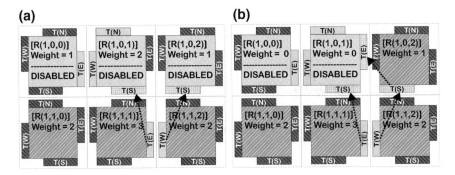

Fig. 6.9 Example of the weight adjustment performed to disable routers' sharing: **a** Before weight update; **b** After weight update

($R(1,0,1)$ and textit$R(1,0,0)$ weights are reduced to zero. As a result, $R(1,0,2)$ can borrow a TSV cluster from $R(1,0,1)$ despite the fact that it originally has a lower weight. The result is shown in Fig. 6.9b where $R(1,0,2)$ vertical connection is re-enabled. If the system want to restart the sharing mechanism, the weights of all routers need to be reinitialized.

6.3.4 Design Optimization

Without adding redundancy, borrowing TSV clusters to work around the defected ones makes some routers to have less than four accessible clusters (e.g., R(1,0,0) in Fig. 6.8d). As a result, the communication of these routers have been disabled. To tackle this problem, the naive solution is using a fault-tolerant routing algorithm to reroute the packets to a neighboring router. This solution may lead to non-minimal routing and congestion in the network. Therefore, we propose *Virtual TSV* to help these routers maintaining the connection without using any fault-tolerant routing algorithm. In the case where the *Virtual TSV* is unable to be performed, we also implement the *Serialization* technique which helps the vertical connection establishing only one or two TSV clusters.

Virtual TSV

When a router is not granted the access to four TSV clusters, it is disabled. However, if the number of nearby TSVs is larger or equal than four, which is enough for maintaining vertical communication, they can be utilized to establish a connection. A possible connection, which requires four TSV clusters, may need clusters belonging to the neighboring routers. If these routers do not use these clusters, the disabled router can borrow them for a short period to establish a communication.

Figure 6.10a shows an example of how *Virtual TSV* works where $R(1,0,1)$ has a defective cluster (T(N)) and borrows a cluster from $R(1,0,0)$. Because $R(1,0,0)$ is unable to find any replacement for the borrowed cluster (T(E)), it is disabled. When $R(1,0,0)$ needs to establish an inter-layer communication, it needs to find at least four TSV clusters. Assuming that $R(1,0,1)$ does not use the borrowed cluster T(E), it is temporarily returned to $R(1,0,0)$. When the packet is completely transmitted, the borrowing cluster is taken back by the router $R(1,0,1)$ again.

On the other hand, Fig. 6.10b shows the case where a disabled router $R(1,0,0)$ temporarily borrows a TSV cluster from a higher weight router $R(1,0,1)$ to establish an inter-layer connection. For selecting a suitable candidate to temporarily borrow, the algorithm shown in Fig. 6.7 is utilized.

Because there is a case where $R(1,0,1)$, which has the higher priority, occupies the TSV for a long transmission time, $R(1,0,0)$ is unable to access the TSV to establish a connection. Moreover, at a high defect rates, $R(1,0,0)$ may not find any suitable candidate for virtual TSV. In order to solve these issues, we adopt the *Serialization* [35] technique to maintain the connection.

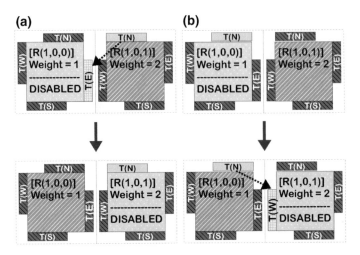

Fig. 6.10 Examples of virtual TSV: **a** return the TSV cluster to the original router; **b** borrow a cluster from a higher weight router

Serialization Technique

Although the *Virtual TSV* can help the disabled router maintaining its vertical connection, there are still two situations where *Virtual TSV* cannot be performed: (a) there are less than four healthy TSV clusters, (b) the candidate TSV cluster is occupied constantly by a higher priority router. In order to solve these cases, we use the *Serialization* technique [35] to maintain the connectivity.

For the serialization, the router needs at least one TSV cluster to maintain its connection. If there is one available cluster, the 1:4 serialization is used, if there are two available clusters, the 1:2 serialization is established. The up and down directions' output of the crossbar is stored in a register and the serialization module transmits flits over the remained clusters. Figure 6.11 shows the vertical interface between two routers using 1:4 serialization. Two serial counters handle the connection by detecting the transmitting flit. This flit is also stored in a buffer in the transmitting router. By increasing the counter's value which selects the multiplexer, the output width is a quarter of the flit size. Because only one TSV cluster is utilized, the controller selects the output by using a demultiplexer.

At the receiving router, the input data will be cached in a register. There are also a demultiplexer and a multiplexer which are controlled by a serial configuration and a serial counter, respectively. When the corresponding counter reaches "11", the whole flit is transmitted to the buffer. For 1:2 serialization, the first half of each flit is cached and when the remainder arrives (counter reach "01"), the whole flit is sent to the buffer.

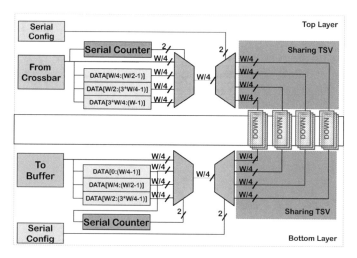

Fig. 6.11 Circuit of 1:4 serialization

6.4   Evaluation Results

The proposed system was designed in Verilog-HDL, synthesized and prototyped with commercial CAD tools. The hardware technology parameters are illustrated in Table 6.2. We use NANGATE 45 nm library [36] and NCSU FreePDK TSV [37]. The system configurations are depicted in Table 6.3.

First, we evaluate the defect rate by inserting faults (defects) into TSV clusters and assess the reliability of the proposed 3D-NoC system. Second, we use both synthetic and realistic traffic patterns as benchmarks to study the performance of the proposed system in comparison to the baseline model [38]. Third, we evaluate the hardware complexity of a single 3D router and compare our system with other proposed approaches [20, 25].

Table 6.2 Technology parameters

Parameter	Value
Technology	Nangate 45 nm [36]
	FreePDK3D45 [37]
Voltage	1.1 V
TSV's size	$4.06\,\mu m \times 4.06\,\mu m$
TSV pitch	$10\,\mu m$
Keep-out zone	$15\,\mu m$

Table 6.3 System configurations

Parameter	Value
# ports	7
Topology	3D mesh
Routing algorithm	Look-ahead routing
Flow control	Stall-go
Forwarding mechanism	Wormhole
Input buffer	4
Flit width	44

6.4.1   Defect-Rate Evaluation

In this section, we provide the impact of the different defect rates. To demonstrate the scalability of the proposed architecture, we set up several layer sizes: 2×2, 4×4, 8×8, 16×16, 32×32, and 64×64. TSVs are grouped in clusters as presented in Sect. 6.2.1. We also vary the TSV-cluster defect rates: from 5 to 50%. Because our technique focuses on the cluster defect, random defects are assumed to be dealt with typical redundancy methods. The position of cluster defects are generated randomly and we perform the proposed algorithms with 100,000 different samples and calculate the average results. We measure the ratio of four types routers in the layer: *Normal* (healthy or corrected), *Virtual* (router with virtual TSV), *Serial* (router using serialization) and *Disabled* (disabled routers). We also compare the obtained results with *"Normal w/o FT"* (Normal without Fault Tolerance), where no fault-tolerance method is used and the router vertical connection having defects is disabled.

As shown in Fig. 6.12, the system mostly operates without disabling any vertical connections with fault-rates under 50%. Thanks to the *Virtual TSV* and *Serialization* techniques, the routers having less than four clusters are still able to work. Even at less than 20% of defect rate, there are less than 10% of serialization connections in all simulated layer sizes. With 50% of defect rate and a 2×2 layer size, the disabled router rate is negligible with about 1.565%. This can be easily dealt using a light-weight fault-tolerant routing algorithm. When the layer size increases to be larger than 8×8, the number of disabled connections is mostly insubstantial. At 50% defect rate, the disabled router ratio is nearly 0.63, 0.50, 0.44 and 0.42% with 8×8, 16×16, 32×32, and 64×64 layer sizes, respectively. However, these defect rates are extremely high; thus, our proposed mechanism can be considered as a highly reliable.

In comparison to the system without fault-tolerant methods, there is a significant improvement in terms of healthy connections, especially at large layer sizes. In Fig. 6.12, the percentage of routers having four healthy TSV clusters is represented by the "Normal w/o FT" curve. At 50% defect rate, the average ratio of normal routers has been improved by 29.83, 186.26, 280.76, 324.42, 346.74, and 257.79% for 2×2, 4×4, 8×8, 16×16, 32×32, and 64×64 layer sizes, respectively. The improvements

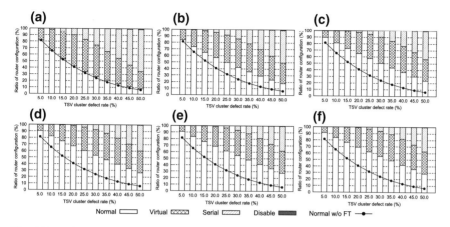

Fig. 6.12 Defect-rate evaluation: **a** Layer size: 2 × 2 (4 routers, 16 TSV clusters); **b** Layer size: 4 × 4 (16 routers, 64 TSV clusters); **c** Layer size: 8 × 8 (64 routers, 256 TSV clusters); **d** Layer size: 16 × 16 (256 routers, 1024 TSV clusters); **e** Layer size: 32 × 32 (1024 routers, 4096 TSV clusters); **f** Layer size: 64 × 64 (4096 routers, 16384 TSV clusters)

are lesser with small layer sizes such as: 2 × 2 or 4 × 4. However, thanks to the Virtual TSV and Serialization, the workable connection rates have nearly reached 100%.

In summary, this evaluation has shown a significant improvement in terms of reliability provided by our proposed mechanism. Thanks to the efficiency of the proposed architecture and algorithm, the system can mostly maintain all vertical connections, even at extremely high defect rate (50%). This evaluation also shows the proposed mechanism ability to remain efficiently scalable. The proposal can be applied from a small layer size (e.g., 2 × 2) to a larger one (e.g., 64 × 64). The evaluation is also performed with a solid number of tests (100,000) which strongly demonstrates the efficiency of the proposed approach. There were some cases where some routers were disabled; however, they can be recovered by simple and lightweight fault-tolerant routing algorithms.

6.4.2   Performance Evaluation

The previous section has proved the reliability of the proposed solution. In this section, we evaluate the system performance under TSV-cluster defects. As we previously mentioned, works in [29, 30] have demonstrated the low utilization rate of the vertical connections; nevertheless, the performance degradation on highly stressed networks has to be investigated. To evaluate the performance of the proposed system and keep fair comparisons to the baseline, we adopted both synthetic and realistic traffic patterns as benchmarks. We selected Transpose [39], Uniform [39], Matrix-multiplication [40], and Hotspot 10% [39] as the synthetic benchmarks. Within these benchmarks, Uniform and Hotspot 10% have the highest stress on the network and

Table 6.4 Simulation configurations

Parameter/System		Value
Network size ($x \times y \times z$)	Matrix	$6 \times 6 \times 3$
	Transpose	$4 \times 4 \times 4$
	Uniform	$4 \times 4 \times 4$
	Hotspot 10%	$4 \times 4 \times 4$
	H.264	$3 \times 3 \times 3$
	VPOD	$3 \times 2 \times 2$
	MWD	$2 \times 2 \times 3$
	PIP	$2 \times 2 \times 2$
Total injected packets	Matrix	1,080
	Transpose	640
	Uniform	8,192
	Hotspot 10%	8,192
	H264	8,400
	VPOD	3,494
	MWD	1,120
	PIP	512
Packet's size	Hotspot 10%	10 flits+10% on hotspot nodes
	Others	10 flits

both Transpose and Matrix-multiplication use vertical connections for all of their connections. For realistic benchmarks, we chose H.264 video encoding system [41], Video Object Plane Decoder (VOPD), Picture In Picture (PIP) and Multiple Window Display (MWD) [42]. These realistic applications are carefully selected to study the performance of the system. Moreover, the network's performance under TSV defects is the focus of these evaluations, the realistic and synthetic benchmarks provide a vast diversity to study the impact of the fault-tolerance. The configurations of these benchmarks are shown in Table 6.4. The packets are injected continuously into the network. In other words, we executed the benchmarks until the saturation point of the network is reached. In order to keep a fair comparison, only TSV defects are injected. This means that the other fault-tolerance mechanisms [43] are disabled to not affect the performance.

6.4.3   Latency Evaluation

In this experiment, we evaluate the performance of the proposed architecture in terms of Average packet Latency (APL) over various benchmark programs and defect rates. The simulation results are shown in Fig. 6.13a. From this graph, we notice that with a 0% of defect rate, the system's tolerance has similar performance in comparison to the baseline system.

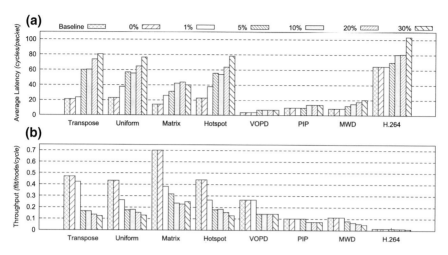

Fig. 6.13 Evaluation result: **a** Average packet latency; **b** Throughput

When we increase the defect rates in the proposed system, it has demonstrated additional impacts on APL. At a 1% fault-rate using Matrix, Uniform, Transpose, and Hotspot 10% benchmarks, the system increases the APL by 83.24, 64.46, 11.30, and 66.55%, respectively. These high impacts are due to the occurrence of bottlenecks inside the network. Because all vertical connections are utilized, Virtual TSV has caused congestions by sharing the TSV between two routers. The serialization is already a bottleneck technique. These bottlenecks effects are even higher at a 30% of defect rate where the APL can be over three times that of the 0% case in the synthetic benchmarks.

With H.264, PIP, MWD and VOPD benchmarks, the APL incrementation are significantly reduced due to the low utilization of TSV. We can observe the identical performance of VOPD benchmark from a 1 to a 30% defect rates. With the PIP benchmark, the system under 1% defect rate has similar performance to 0% thank to the optimization process which disables the unused clusters. With the MWD and H.264 benchmarks, the impact on APL is gradually increased when increasing the defect rate. Even at a 30% of defect rate, the APL values of MWD and H.264 are increased by 129.91 and 60.04%, respectively. Because there is no optimized routing technique for these benchmarks, the bottleneck effect is expected to happen.

Although there are significant impacts in latency, the system has proven to work without major issues in all benchmarks.

6.4.4 Throughput Evaluation

Figure 6.13b depicts the throughput evaluation with different benchmarks. At 0% defect rate, the proposed system's throughput is similar to that of the baseline. When defects are injected into the system, we can observe some degradation in throughput caused by the bottleneck effects on the system. Similar to APL, the throughput degradation on realistic traffic benchmarks (VOPD, H.264, MWD and PIP) are significantly better than the synthetic ones. The system at a 20% defect rate provides a decreased throughput by 71.17, 64.36, 67.44 and 64.37% for Transpose, Uniform, Matrix and Hotspot 10%, respectively. At the same defect rate, VOPD, MWD, PIP and H.264 have 46.03, 50.04, 28.17 and 19.79% of throughput degradation. This lower impact is caused by the low utilization of vertical connection rate and the optimization process. The throughput of realistic benchmarks are naturally smaller than the synthetic ones because of the specific tasks order of execution that was observed in the task graphs [41, 42].

Although there is a considerable degradation in the throughput evaluation, the system still maintains over 0.1 *flit/node/cycle* in the highly stressed benchmarks, even at extremely high defect rates.

6.4.5   Router Hardware Complexity

Table 6.5 illustrates the hardware complexity results of the proposed router in terms of area, power (static, dynamic, and total), and speed. In comparison to the router in which we implement the proposed techniques, the area, and power consumption have increased by 30.42 and 18.66%, respectively. The maximum speed has also slightly decreased by 12.37%. In comparison to the baseline model, the proposed system almost doubles the area cost and power consumption while decreasing the maximum frequency by about 50%. However, the TSV sharing and Serialization modules incur reasonable area and power consumption overheads which are 47.99 and 38.89% in comparison to the baseline router, respectively. Here, the TSV Sharing module handles the sharing algorithm and the *Virtual TSV* process and the Serialization module helps the router communicate in *Serialization* mode.

The layout of a layer is shown in Fig. 6.14 where the sharing TSV areas are depicted by the red boxes. As shown in Sect. 6.2.1.2, the TSV sharing area consists of eight clusters. For each port, *R(1,1,1)* can access *T(E)* of *R(1,1,0)* and *R(1,1,0)* can access *T(W)* of *R(1,1,1)*. By placing the shared cluster areas between two routers, we can ensure a small extra wire delay for rerouting.

Table 6.5 Hardware complexity of a single router

Model		Area (μm^2)	Power (mW)			Speed (Mhz)
			Static	Dynamic	Total	
Baseline router [38]		18,873	5.1229	0.9429	6.0658	925.28
Proposal	Router	29,780	10.017	2.2574	12.3144	613.50
	Serialization	3,318	0.9877	0.2807	1.2684	–
	TSV sharing	5,740	0.7863	0.2892	1.0300	–
	Total	38,838	11.7910	2.8273	14.6128	537.63

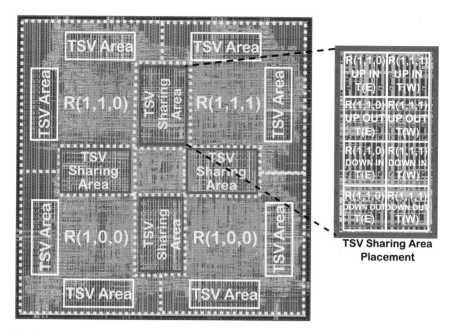

Fig. 6.14 Single layer layout illustrating the TSV sharing areas (*red boxes*). The layout size is 865 μm × 865 μm

6.4.6 Comparison

In order to understand the efficiency of the proposed approach, we compare it with existing solutions as shown in Table 6.6. Here, we analyze our proposal with a network size of 4 × 4 × 4. Because the router and its TSV clusters structure are identical, similar results can be obtained with the others network sizes. *TSV Grouping* [25] optimized the configuration of redundancy to deal with TSV-cluster defects. *TSV Network* [20] established TSVs into networks which allow routing from defected TSVs to redundant ones. We select the best results on these two works [20, 25] for the comparison. From this table, we can see that the average area of our proposal is 151.47 μm² per TSV and, for a TSV size of 10 μm × 10 μm, the area overhead is about 51.47%. The *TSV Network* [20] has similar value for 4:2 configuration (four original TSVs and two redundant TSVs). With 8:4 configuration, *TSV Grouping* also obtained an average area of 151.86 μm².

On the other hand, the other configurations obtained lower area overheads. Nevertheless, we have to note that our arbiter not only consists of the rerouting circuit (similar to the multiplexers in *TSV Network* and *TSV Grouping*); but, also includes an online adaptive algorithm designed in hardware, in addition to the *Virtual TSV* and *Serialization* techniques. Both *TSV Grouping* and *TSV Network* have to require additional dedicated circuitry to recover from the cluster defects.

Table 6.6 Comparison results between the proposed approach and the existing works

Model	TSV network [20]				TSV grouping [25]		This work
Technology	65 nm				N/A		45 nm
#TSV	1000				6000		8448
Configuration	4:2	8:2	4 × 4 : 8	16 × 16 : 32	4:4	8:4	11 × 4 × 4 : 0
#Spare TSV	512	256	512	128	6000	3000	0
45 nm Arbiter area (μm^2)	372[b]	744[b]	1,116[b]	1,116[b]	11,160[a]	11,160[a]	434,784[c]
Average area/TSV (μm^2)	151.572	126.244	152.316	128.03	113.916	151.86	151.47
Reliability	100%	99%	100%	100%	100%	100%	98.11%
Fault assumption	$(\delta_{TSV} = 0.01\%, \alpha = 2)^d$				$(\delta_{TSV} = 1\%, \alpha = 2)^d$		$(\delta_{cluster} = 50\%)^d$

[a]The authors use 2:1 multiplexers [25]. For comparison, we use the area cost of multiplexer from Nangate 45 nm [36] (MUX2_X1: 0.186 μm^2)
[b]The authors use 1–3 multiplexers [20] which consists of two MUX2_X1 multiplexers ($2 \times 0.186 \, \mu m^2$ [36])
[c]For fair comparisons, our arbiter only consists of the TSV sharing and serialization modules as shown in Table 6.5
[d]δ: defect rate. α: parameter of Poisson distribution [20, 25]

In terms of reliability, the proposed approach has proven its high resiliency, as previously shown in Sect. 6.4.1. *TSV Grouping* demonstrated a 100% of yield rate under a defect rate of 1% and *TSV Network* obtained nearly 100% in the most cases. However, their approaches are different than our scheme, where they add redundancy to correct the defect TSVs. As a result, if the number of defected TSVs is larger than the number of redundant ones, they are unable to recover from the defected clusters. On the other hand, our technique can significantly improve the reliability by providing 98.11% of workable routers at 50% of defected TSV clusters. Moreover, at the low defect rates (e.g., under 5%), our proposal also ensures 100% of working connection and demonstrates small performance degradations in the realistic traffic pattern benchmarks. Even with disabled vertical connections, the reliability of our system can also be improved (i.e., covering the remaining 1.89%) by using a lightweight fault-tolerant routing which would have a negligible impact on the area overhead.

6.5 Chapter Summary

This chapter presented an adaptive and scalable sharing methodology for TSVs in 3D-NoC systems to deal with the TSV-cluster defects. The results have proven the system ability to provide high reliability that can reach up to 346.74% increase in functional routers. Moreover, the proposed approach can correctly work with a reasonable degradation, even under a 30% of defect rate. The hardware complexity has shown a small overhead in terms of area cost (30.42%), power consumption (18.66%), and maximum frequency (12.37%) of router's logic. Since no TSV redundancy is not required in the proposed architecture and algorithm, we show that it is possible to provide a highly reliable system while maintaining the overhead reasonable.

References

1. U. Kang, H.-J. Chung, S. Heo, S.-H. Ahn, H. Lee, S.-H. Cha, J. Ahn, D. Kwon, J. Kim, J.-W. Lee, et al., 8Gb 3D DDR3 DRAM using through-silicon-via technology, in *IEEE International Solid-State Circuits Conference-Digest of Technical Papers* (ISSCC)
2. J. Ahn et al., 7.1 A 1/4-inch 8Mpixel CMOS image sensor with 3D backside-illuminated 1.12 um pixel with front-side deep-trench isolation and vertical transfer gate, in *2014 IEEE International Solid-State Circuits Conference Digest of Technical Papers (ISSCC)* (2014), pp. 124–125
3. V. Suntharalingam, R. Berger, S. Clark, J. Knecht, A. Messier, K. Newcomb, D. Rathman, R. Slattery, A. Soares, C. Stevenson, et al., A 4-side tileable back illuminated 3d-integrated mpixel cmos image sensor, in *IEEE International of Solid-State Circuits Conference-Digest of Technical Papers, 2009. ISSCC 2009*, (IEEE, New York, 2009), pp. 38–39
4. H. Yoshikawa, A. Kawasaki, T. Iiduka, Y. Nishimura, K. Tanida, K. Akiyama, M. Sekiguchi, M. Matsuo, S. Fukuchi, K. Takahashi, Chip scale camera module (CSCM) using through-

silicon-via (TSV), in *IEEE International Solid-State Circuits Conference–Digest of Technical Papers* (2009), pp. 476–477, 477a

5. K. Ishida, T. Yasufuku, S. Miyamoto, H. Nakai, M. Takamiya, T. Sakurai, K. Takeuchi, A 1.8 V 30nJ adaptive program-voltage (20V) generator for 3D-integrated NAND flash SSD, in *IEEE International of Solid-State Circuits Conference-Digest of Technical Papers, 2009. ISSCC 2009*, (IEEE, New York, 2009), pp. 238–239

6. M. Saen, K. Osada, Y. Okuma, K. Niitsu, Y. Shimazaki, Y. Sugimori, Y. Kohama, K. Kasuga, I. Nonomura, N. Irie et al., 3-d system integration of processor and multi-stacked srams using inductive-coupling link. IEEE J. Solid-State Circuits **45**(4), 856–862 (2010)

7. M. Karnezos, 3d packaging: Where all technologies come together, in *Electronics Manufacturing Technology Symposium, 2004. IEEE/CPMT/SEMI 29th International* (IEEE, New York, 2004), pp. 64–67

8. J. Miettinen, M. Mantysalo, K. Kaija, E. Ristolainen, System design issues for 3d system-in-package (sip), in *Electronic Components and Technology Conference, 2004. Proceedings. 54th*, vol. 1 (IEEE, New York, 2004), pp. 610–615

9. K. Banerjee, S.J. Souri, P. Kapur, K.C. Saraswat, 3-D ICs: a novel chip design for improving deep-submicrometer interconnect performance and systems-on-chip integration. Proc. IEEE **89**(5), 602–633 (2001)

10. E. Culurciello, A.G. Andreou, Capacitive inter-chip data and power transfer for 3-d vlsi. IEEE Trans.Circuits Syst. II: Express Briefs **53**(12), 1348–1352 (2006)

11. W.R. Davis, J. Wilson, S. Mick, J. Xu, H. Hua, C. Mineo, A.M. Sule, M. Steer, P.D. Franzon, Demystifying 3d ics: the pros and cons of going vertical. IEEE Des. Test Comput. **22**(6), 498–510 (2005)

12. A.B. Ahmed, A. Ben Abdallah, Adaptive fault-tolerant architecture and routing algorithm for reliable many-core 3D-NoC systems. J. Parallel Distrib. Comput. **9394**(7), 30–43 (2016)

13. A.B. Ahmed, A. Ben Abdallah, Architecture and design of high-throughput, low-latency, and fault-tolerant routing algorithm for 3d-network-on-chip (3d-noc). J. Supercomput. **66**(3), 1507–1532 (2013)

14. A.B. Ahmed, A. Ben Abdallah, K. Kuroda, Architecture and design of efficient 3d network-on-chip (3D NoC) for custom multicore soc, in *IEEE Proceedings of BWCCA-2010* (2010)

15. K.N. Dang, M. Meyer, Y. Okuyama, A. Ben Abdallah, A low-overhead soft-hard fault tolerant architecture, design and management scheme for reliable high-performance many-core 3D-NoC systems. Supercomputer **73**, 2705–2729 (2017)

16. J.U. Knickerbocker, P.S. Andry, B. Dang, R.R. Horton, M.J. Interrante, C.S. Patel, R.J. Polastre, K. Sakuma, R. Sirdeshmukh, E.J. Sprogis et al., Three-dimensional silicon integration. IBM J. Res. Dev. **52**(6), 553–569 (2008)

17. T. Zhang, Y. Zhan, S. Sapatnekar, Temperature-aware routing in 3D ICs, in *Asia and South Pacific Conference on Design Automation* (2006), pp. 309–314

18. M. Cho, C. Liu, D.H. Kim, S.K. Lim, S. Mukhopadhyay, Design method and test structure to characterize and repair TSV defect induced signal degradation in 3D system, in *Proceedings of the International Conference on Computer-Aided Design*, (IEEE Press, New York, 2010), pp. 694–697

19. M. Laisne, K. Arabi, T. Petrov, Systems and methods utilizing redundancy in semiconductor chip interconnects, US Patent 8,384,417, 2013

20. L. Jiang, F. Ye, Q. Xu, K. Chakrabarty, B. Eklow, On effective and efficient in-field TSV repair for stacked 3D ICs, in *Proceedings of the 50th Annual Design Automation Conference*, (ACM, 2013) p. 74

21. I. Loi, S. Mitra, T.H. Lee, S. Fujita, L. Benini, A low-overhead fault tolerance scheme for TSV-based 3D network on chip links, in *Proceedings of the 2008 IEEE/ACM International Conference on Computer-Aided Design*, (IEEE Press, New York, 2008), pp. 598–602

22. Y. Zhao, S. Khursheed, B.M. Al-Hashimi, Online fault tolerance technique for TSV-based 3-D-IC. IEEE Trans. Very Large Scale Integr. (VLSI) Syst. **23**(8), 1567–1571 (2015)

23. A.-C. Hsieh, T. Hwang, TSV redundancy: architecture and design issues in 3-D IC. IEEE Trans. Very Large Scale Integr. (VLSI) Syst. **20**(4), 711–722 (2012)

24. F. Ye, K. Chakrabarty, TSV open defects in 3D integrated circuits: characterization, test, and optimal spare allocation, in *Proceedings of the 49th Annual Design Automation Conference*, (ACM, 2012), pp. 1024–1030
25. Y. Zhao, S. Khursheed, B.M. Al-Hashimi, Cost-effective TSV grouping for yield improvement of 3D-ICs, in *Asian Test Symposium (ATS)*, (IEEE, New York, 2011), pp. 201–206
26. D. Bertozzi, L. Benini, G. De Micheli, Error control schemes for on-chip communication links: the energy-reliability tradeoff. IEEE Trans. Comput. Aided Des. Integr. Circuits Syst. **24**(6), 818–831 (2005a)
27. A.B. Ahmed, A. Ben Abdallah, Architecture and design of high-throughput, low-latency, and fault-tolerant routing algorithm for 3D-network-on-chip (3D-NoC). J. Supercomput. **66**(3), 1507–1532 (2013)
28. K.C.J. Chen, C.H. Chao, A.Y.A. Wu, Thermal-aware 3D network-on-chip (3D NoC) designs: routing algorithms and thermal managements. IEEE Circuits Syst. Mag. **15**(4), 45–69 (2015)
29. Y.J. Hwang, J.H. Lee, T.H. Han, 3d network-on-chip system communication using minimum number of tsvs, in *ICT Convergence (ICTC), 2011 International Conference on*, (IEEE, New York, 2011), pp. 517–522
30. A. Kologeski, C. Concatto, D. Matos, D. Grehs, T. Motta, F. Almeida, F.L. Kastensmidt, A. Susin, R. Reis, Combining fault tolerance and serialization effort to improve yield in 3d networks-on-chip, in *2013 IEEE 20th International Conference on Electronics, Circuits, and Systems (ICECS)* (2013), pp. 125–128
31. Y.-J. Huang, J.-F. Li, Built-in self-repair scheme for the TSVs in 3-D ICs. IEEE Trans. Comput. Aided Des. Integr. Circuits Syst. **31**(10), 1600–1613 (2012)
32. M. Tsai, A. Klooz, A. Leonard, J. Appel, P. Franzon. Through silicon via (TSV) defect/pinhole self test circuit for 3D-IC, in *IEEE International Conference on 3D System Integration*, (IEEE, New York, 2009), pp. 1–8
33. M. Palesi, R. Holsmark, S. Kumar, V. Catania, Application specific routing algorithms for networks on chip. IEEE Trans. Parallel Distrib. Syst. **20**(3), 316–330 (2009)
34. Z. Qian and C. Y. Tsui. A thermal-aware application specific routing algorithm for network-on-chip design, in *16th Asia and South Pacific Design Automation Conference (ASP-DAC 2011)* (2011), pp. 449–454
35. Y. Ghidini, M. Moreira, L. Brahm, T. Webber, N. Calazans, C. Marcon, Lasio 3D NoC vertical links serialization: Evaluation of latency and buffer occupancy, in *26th Symposium on Integrated Circuits and Systems Design (SBCCI)* (2013), pp. 1–6
36. NanGate Inc. Nangate open cell library 45 nm, (2016)
37. NCSU Electronic Design Automation. FreePDK3D45 3D-IC process design kit, (2016)
38. A.B. Ahmed, A. Ben Abdallah, LA-XYZ: low latency, high throughput look-ahead routing algorithm for 3D network-on-chip (3D-NoC) architecture, in *IEEE 6th International Symposium on Embedded Multicore Socs (MCSoC)*, (IEEE, New York, September 2012), pp. 167–174
39. W.J. Dally, B.P. Towles, *Principles and Practices of Interconnection Networks* (Elsevier, Beijing, 2004)
40. P. Chen, K. Dai, D. Wu, J. Rao, X. Zou, The parallel algorithm implementation of matrix multiplication based on ESCA, in *IEEE Asia Pacific Conference on Circuits and Systems (APCCAS)*, (IEEE, New York, 2010), pp. 1091–1094
41. A.-M. Rahmani, K.R. Vaddina, K. Latif, P. Liljeberg, J. Plosila, H. Tenhunen, High-performance and fault-tolerant 3D noc-bus hybrid architecture using arb-net-based adaptive monitoring platform. IEEE Trans. Comput. **63**(3), 734–747 (2014)
42. D. Bertozzi, A. Jalabert, S. Murali, R. Tamhankar, S. Stergiou, L. Benini, G. De Micheli, NoC synthesis flow for customized domain specific multiprocessor systems-on-chip. IEEE Trans. Parallel Distrib. Syst. **16**(2), 113–129 (2005)
43. K.N. Dang, M. Meyer, Y. Okuyama, A. Ben Abdallah, A low-overhead soft–hard fault-tolerant architecture, design and management scheme for reliable high-performance many-core 3D-NoC systems. J. Supercomput. **73**(6), 1–25 (2017)

Chapter 7
Parallelizing Compiler for Single and Multicore Computing

Abstract To overcome challenges from high power densities and thermal hot spots in microprocessors, multicore computing platforms have emerged as the ubiquitous computing platform from servers to embedded systems. But, providing multiple cores does not directly translate into increased performance for most applications. The burden is placed on software developers to find and exploit coarse-grain parallelism to effectively make use of the abundance of computing resources provided by the systems. With the rise of multicore systems and many-core processors, concurrency becomes a major issue in the daily life of a programmer. Thus, compiler and software development tools will be critical to help programmers create high-performance software. This chapter covers software issues of a so-called parallelizing queue compiler targeted for future single- and multicore embedded systems.

7.1 Introduction

A compiler is a program that translates one computer language into another language (target language). Most of the compilers translate a high-level programming language into machine language program or also called the object code. The goal of high-level programming languages is to hide the details of the microprocessor in a set of abstract, easy to use concepts to make complex programming simpler. Sophisticated programs such as operating systems and modern applications rely on high-level programming languages to facilitate their development, to reduce implementation time, and to avoid error-prone assembly programming. Compilers are a very important layer in the computer systems stack as they translate application code into machine code with comparable, or better, performance than hand-coded assembly. Although compiler technology is well understood for parallelizing programs for single-core processors, the introduction of many cores has brought major challenges for the compiler designers.

© Springer Nature Singapore Pte Ltd. 2017
A. Ben Abdallah, *Advanced Multicore Systems-On-Chip*,
DOI 10.1007/978-981-10-6092-2_7

7.1.1 *Instruction Level Parallelism*

Instruction level parallelism (ILP) is the key to improve the performance of modern architectures. ILP allows the instructions of a sequential program to be executed in parallel on multiple data paths and functional units. Data and control independent instructions determine the groups of instructions that can be issued together while keeping the program correctness [1].

A good scheduling is crucial to achieve high performance. An effective scheduling for the exploitation of ILP depends greatly on two factors: the processor features, and the compiler techniques. In superscalar processors, the compiler exposes ILP by rearranging instructions. However, the final schedule is decided at runtime by the hardware [2]. In VLIW machines, the scheduling is decided at compile time by aggressive static scheduling techniques [1, 3].

Sophisticated compiler optimizations have been developed to expose high amounts of ILP in loop regions [4], where many scientific and multimedia programs spend most of their execution time. The purpose of some loop transformations such as loop unrolling is to enlarge basic blocks by combining instructions called in multiple iterations to a single iteration. A popular loop scheduling technique is modulo scheduling [5, 6] where the iterations of a loop are parallelized in such a way that a new iteration initiates before the previous iteration has completed execution.

These static scheduling algorithms improve greatly the performance of the applications at the cost of increasing the register pressure [7]. When the schedule requires more registers than those available in the processor, the compiler must insert spill code to fit the application in the available number of architected registers [8]. Many high-performance architectures born in the last decade [9–11] were designed on the assumption that applications could not make effective use of more than 32 registers [12]. Recent studies have shown that the register requirements for the same kind of applications using the current compiler technology demand more than 64 registers [13].

High ILP register requirements have direct impact in the processor performance as a large number of registers need to be accessed concurrently. The number of ports to access the register file affects the access time and the power consumption. In order to maintain clock speed and low power consumption, high-performance embedded and digital signal processors have implemented partitioned register banks [14] instead of a large monolithic register file. Several software solutions for the compiler have been proposed to reduce the register requirements of modulo schedules [15], and other studies have focused on the compilation issues for partitioned register files [16, 17]. A hardware/compiler technique to alleviate register pressure is to provide more registers than allowed by the instruction encoding. In [18, 19], the usage of queue register files has been proposed to store the live variables in a software pipelined loop schedule while minimizing the pressure on the architected registers. The work in [20] proposes the use of register windows to give the illusion of a large register file without affecting the instruction set bits.

An alternative to hide the registers from the instruction set encoding is by using a queue machine. A queue machine uses a first-in first-out structure, called the operand queue, as the intermediate storage location for computations. Instructions read and write the operand queue implicitly. Not having explicit operands in the instructions make instructions short improving code density. Also, false dependencies disappear from programs eliminating the need for register renaming logic that reduces circuitry and improves power consumption [21].

Queue computers have been studied in several works. Bruno [22] investigated the possibility of evaluating expression trees and highlighted the problems of evaluating directed acyclic graphs (DAG) in an abstract queue machine.

In [23], Okamoto presented some design issues of a superscalar queue machine. Schmit et al. [24] use a queue machine as the execution layer for reconfigurable hardware. They transform the program's data flow graph (DFG) into a spatial representation that can be executed in a simple queue machine. This transformation inserts extra special instructions to guarantee correct execution by allowing every variable to be produced and consumed only once. Their experiments show that the execution of programs in their queue machine has the potential of exploiting high levels of parallelism while keeping code size less than a RISC instruction set.

In [25, 26], a 32-bit QueueCore processor with a 16-bit instruction set format was designed. The approach is to allow variables to be produced only once but can be consumed multiple times. We sacrifice some bits in the instruction set for an offset reference to indicate the relative location of a variable to be reused. The goal is to allow DAGs to be executed without transformations that increase the instruction count while keeping reduced instructions that generate dense programs.

Ideas about compiling for queue machines have been discussed in the previous work in an abstract way. Some problems have been clearly identified but no algorithms have been proposed. Before, we explored the possibility of using a retargettable code generator for register machines to map register code into the queue computation model [27]. The resulting compiler mapped the operand queue in terms of a large number general purpose registers in the machine description file that is used by the code generator in order to avoid spill code. This approach led to complex algorithms to map register programs into queue programs, excessively long programs, poor parallelism, and poor code quality.

This chapter presents a code generation scheme implemented in a compiler for the QueueCore processor. The compiler generates assembly code from C programs and is suitable for single-core and multicore platforms. The queue compiler exposes natural ILP from the input programs to the QueueCore processor. Experimental results show that the compiler can extract more parallelism for the QueueCore than an ILP compiler for a RISC machine, and also generate programs with lower code size.

7.1.2 Queue Computation Model

A queue-based computer employs a first-in first-out (FIFO) queue to evaluate expressions [25, 28, 29]. To avoid high-latency memory accesses, the queue is implemented with high-speed registers arranged and accessed in a special manner. The physical implementation of the queue is called the Queue Register File. Reading operation from the queue is done always through the head of the queue, and writing operation is done always through the tail of the queue. Therefore, the hardware must provide two pointers to track the head and tail of the queue. Such pointers are implemented as special registers, QH and QT to track the head and the tail positions of the queue. The queue computation model (QCM) is the set of rules and conventions that allow programs to be executed in a queue processor.

7.2 Parallel Queue Compiler

The queue computation model (QCM) is the abstract definition of a computer that uses a first-in first-out data structure as the storage space to perform operations. Elements are inserted, or enqueued, through a write pointer named QTthat references the rear of the queue. And elements are removed, or dequeued, through a read pointer named QHthat references the head of the queue.

7.2.1 Queue Processor Overview

The QueueCore is a 32-bit processor with a 16-bit wide producer order QCM instruction set architecture based on the produced order parallel QCM [26, 28, 30–32]. The instruction format reserves 8-bit for the opcode and 8-bit for the operand. The operand field is used in binary operations to specify the offset reference value with respect of QHfrom which the second source operand is dequeued, $QH-N$. Unary operations have the freedom to dequeued their only source operand from $QH-N$. Memory operations use the operand field to represent the offset and base register, or immediate value. For cases when 8-bit is not enough to represent an immediate value or an offset for a memory instruction, a special instruction named "covop" is inserted before the conflicting memory instruction. The "covop" instruction extends the operand field of the following instruction.

QueueCore defines a set of specific purpose registers available to the programmer to be used as the frame pointer register ($fp), stack pointer register ($sp), and return address register ($ra). Frame pointer register serves as a base register to access local variables, incoming parameters, and saved registers. Stack pointer register is used as the base address for outgoing parameters to other functions.

7.2.2 Compiling for One-Offset QueueCore Instruction Set

The instruction sequence to correctly evaluate a given expression is generated from a level-order traversal of the expressions' parse tree [22]. A level-order traversal visits all the nodes in the parse tree from left to right starting from the deepest level towards the root as shown in Fig. 7.1a.

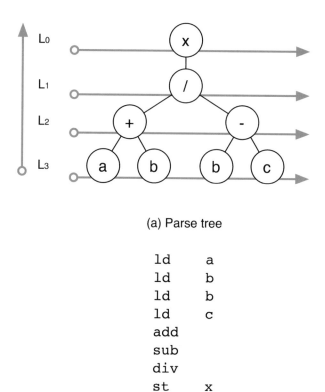

(a) Parse tree

```
ld      a
ld      b
ld      b
ld      c
add
sub
div
st      x
```

(b) Instruction Sequence

QSTATE	Level	Consume	Produce
1	L3	0	4
2	L2	4	2
3	L1	2	1
4	L0	1	0

(c) QSTATEs

Fig. 7.1 Instruction sequence generation from the parse tree of expression $x = \frac{a+b}{b-c}$

The generated instruction sequence is shown in Fig. 7.1b. All nodes in every level are independent from each other and can be processed in parallel. Every node may consume and produce data. For example, a load operation produces one datum and consumes none, a binary operation consumes two data and produces one. A QSTATE is the relationship between all the nodes in a level that can be processed in parallel and the total number of data consumed and produced by the operations in that level. Figure 7.1c shows the production and consumption degrees of the QSTATEs for the sample expression.

Although the instruction sequence from a directed acyclic graph (DAG) is obtained also from a level-order traversal, there are some cases where the basic rules of enqueueing and dequeueing are not enough to guarantee correctness of the program [22]. Figure 7.2a shows the evaluation of an expression's DAG that leads to incorrect results. In Fig. 7.2c, notice that at QSTATE 1 there are three operands produced, and at QSTATE 2 the operations consume four operands. The add operation in Fig. 7.2b consumes two operands, a, b, and produces one, the result of the addition $a + b$. The sub operation consumes two operands that should be b, c, instead it consumes operands c, $a + b$.

In our previous work [26], we have proposed a solution for this problem. We give flexibility to the dequeueing rule to get operands from any location in the operand queue. In other words, we allow operands to be consumed multiple times. The desired operand's location is relative to the head of the queue and it is specified in the instruction as an offset reference, $QH-N$. As the enqueueing rule, production of data, remains fixed at QT, we name this model the Producer Order Queue Computation Model.

Figure 7.1 shows the code for this model that solves the problems in Fig. 7.2. Notice that add, sub, div instructions have offset references that indicate the place relative to QH where the operands should be taken. The "sub -1, 0" instruction now takes operand b from $QH-1$, and operand c from QH itself, QH+0. We name the code for this model P-Code. This nontraditional computation model requires new compiler support to statically determine the value of the offset references.

Correct evaluation of binary instructions whose both source operands are away from QH using QueueCore's one operand instruction set is not possible. To ensure correct evaluation of this case, a special instruction has been implemented in the processor. The dup instruction takes a variable in the operand queue and places a copy in QT. The compiler is responsible for placing dup instructions to guarantee that binary instructions will have their first operand available always at QH, and the second operand may be taken from an arbitrary position in the operand queue by using QueueCore's one operand instruction set. Let the expression $x = -a/(a + a)$ be evaluated using QueueCore's one offset instruction set, its DAG is shown in Fig. 7.2a. Notice that the level L_3 produces only one operand, a, that is consumed by the following instruction, neg. The add instruction is constrained to take its first source operand directly from QH, and its second operand has freedom to be

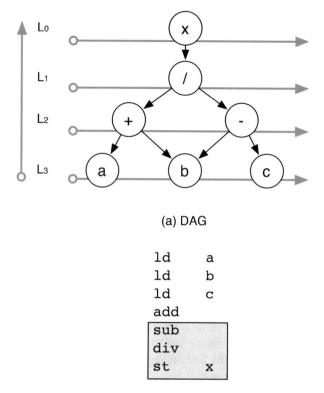

(a) DAG

```
ld      a
ld      b
ld      c
add
sub
div
st      x
```

(b) Instruction Sequence

QSTATE	Level	Consume	Produce
1	L3	0	3
2	L2	4	2
3	L1	2	1
4	L0	1	0

(c) QSTATEs

Fig. 7.2 Instruction sequence generation from DAG of expression $x = \frac{a+b}{b-c}$

taken from $QH-N$. For this case, the dup instruction is inserted to make a copy of a available as the first source operand of instruction add as shown with the dashed line in Fig. 7.2b. Notice that level L_3 in Fig. 7.2b produces two data instead of one. The instruction sequence using QueueCore's one offset instruction set is shown in Fig. 7.2c. This mechanism allows safe evaluation of binary operations in a DAG using one offset instruction set at the cost of the insertion of dup instructions. The QueueCore's instruction set format was decided from our design space exploration

[32]. We found that binary operations that require the insertion of dup instructions are rare in program DAGs. We believe that one operand instruction set is a good design to keep a balance between compact instructions and program requirements.

7.3 Parallelizing Compiler Framework

There are three tasks, the parallelizing queue compiler must do that to make it different from traditional compilers for register machines:

(1) constrain all instructions to have at most one offset reference,
(2) compute offset reference values, and
(3) schedule the program expressions in level-order manner.

We developed a C compiler for the QueueCore that uses GCC's 4.0.2 front end and middle end. The C program is transformed into abstract syntax tree (AST) by the front-end. Then the middle end converts the ASTs into a language and machine independent format called GIMPLE [33]. A set of tree transformations and optimizations to remove redundant code and substitute sequences of code with more efficient sequences is optionally available from the GCC's middle end for this representation. Although these optimizations are available in our compiler, until this point, our primary goal was to develop the basic compiler infrastructure for the QueueCore and we have not validated the results and correctness of programs compiled with these optimizations enabled. We wrote a custom back end that takes GIMPLE intermediate representation and generates assembly code for the QueueCore processor. Figure 7.3 shows the phases and intermediate representations of the queue compiler infrastructure. The uniqueness of our compiler is from the One-offset code generation algorithm implemented as the first and second phases in the back end. This algorithm transforms the data flow graph to assure that the program can be executed using a one-offset queue instruction set. The algorithm then statically determines the offset values for all instructions by measuring the distance of QH relative position with respect to each instruction. Each offset value is computed once and remains the same until the final assembly code is generated. The third phase of the back end converts our middle-level intermediate representation into a linear one-operand low-level intermediate code, and at the same time, schedules the program in a level-order manner. The linear low-level code facilitates the extraction of natural ILP done by the fourth phase. Finally, the fifth phase converts the low-level representation of the program into assembly code for the QueueCore. The following subsections describe in detail the phases, the algorithms, and the intermediate representations utilized by our queue compiler to generate assembly code from any C program.

Fig. 7.3 Parallelizing
compiler infrastructure

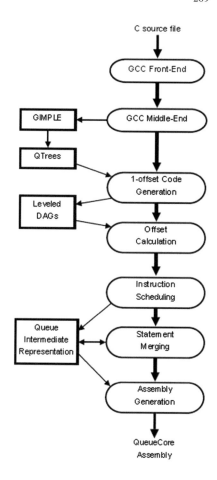

7.3.1   One-Offset P-Code Generation Phase

GIMPLE is a three address code intermediate representation used by GCC's middle end to perform optimizations. Three address code is a popular intermediate representation in compilers that expresses well the instructions for a register machine, but fails to express instructions for the queue computation model. The first task of our back end is to expand the GIMPLE representation into QTrees. QTrees are ASTs without limitation in the number of operands and operations.

GIMPLE's high-level constructs for arrays, pointers, structures, unions, subroutine calls are expressed in simpler GIMPLE constructs to match the instructions available in a generic queue hardware.

The task of the first phase of our back end, one-offset P-Code Generation, is to constrain the binary instructions in the program to have at most one offset reference. This phase detects the cases when dup instructions need to be inserted and it determines the correct place. The code generator takes as input QTrees and generates leveled directed acyclic graphs (LDAGs) as output. A leveled DAG is a data structure that binds the nodes in a DAG to levels [34]. We chose LDAGs as data structure to model the data dependencies between instructions and QSTATEs.

The algorithm works in two stages. The first stage converts QTrees to LDAGs augmented with ghost nodes. A ghost node is a node without operation that serves as a mark for the algorithm. The second stage takes the augmented LDAGs and removes all ghost nodes by deciding whether a ghost node becomes a dup instruction or is removed.

7.3.1.1 Augmented LDAG Construction

QTrees are transformed into LDAGs by a post-order depth-first recursive traversal over the QTree. All nodes are recorded in a lookup table when they first appear and are created in the corresponding level of the LDAG together with its edge to the parent node. Two restrictions are imposed over the LDAGs for the one-offset P-Code QCM.

Definition 7.3.1 A level is an ordered list of elements with at least one element.

Definition 7.3.2 The sink of an edge must be always in a deeper or same level than its source.

Definition 7.3.3 An edge to a ghost node spans only one level.

When an operand is found in the lookup table, the Definition 7.3.2 must be kept. Line 5 in Algorithm 7.1 is reached when the operand is found in the lookup table and it has a shallower level (closer to the root) than the new level. The function dag_ghost_move_node() moves the operand to the new level, updates the lookup table, converts the old node into a ghost node, and creates an edge from the ghost node to the newly created node.

The function insert_ghost_same_level() in Line 8 is reached when the level of the operand in the lookup table is the same to the new level. This function creates a new ghost node in the new level, makes an edge from the parent node to the ghost node, and an edge from the ghost node to the element matched in the lookup table. These two functions build LDAGs augmented with ghost nodes that obey Definitions 7.3.2 and 7.3.3.

Algorithm 7.1 dag_levelize_ghost (tree t, level)

1: nextlevel $\Leftarrow$ level + 1
2: match $\Leftarrow$ lookup (t)
3: **if** match $\neq$ null **then**
4:   **if** match.level < nextlevel **then**
5:     relink $\Leftarrow$ dag_ghost_move_node (nextlevel, t, match)
6:     **return** relink
7:   **else if** match.level = lookup (t) **then**
8:     relink $\Leftarrow$ insert_ghost_same_level (nextlevel, match)
9:     **return** relink
10:   **else**
11:       **return** match
12:   **end if**
13: **end if**
14: /* Insert the node to a new level or existing one */
15: **if** nextlevel > get_Last_Level() **then**
16:   new $\Leftarrow$ make_new_level (t, nextlevel)
17:   record (new)
18: **else**
19:   new $\Leftarrow$ append_to_level (t, nextlevel)
20:   record (new)
21: **end if**
22: /* Post-Order Depth First Recursion */
23: **if** t is binary operation **then**
24:   lhs $\Leftarrow$ dag_levelize_ghost (t.left, nextlevel)
25:   make_edge (new, lhs)
26:   rhs $\Leftarrow$ dag_levelize_ghost (t.right, nextlevel)
27:   make_edge (new, rhs)
28: **else if** t is unary operation **then**
29:   child $\Leftarrow$ dag_levelize_ghost (t.child, nextlevel)
30:   make_edge (new, child)
31: **end if**
32: **return** new

7.3.1.2 dup Instruction Assignment and Ghost Nodes Elimination

The second and final stage of the one-offset P-Code generation algorithm takes the augmented LDAG and decides what ghost nodes are assigned to be a dup node or eliminated from the LDAG. The only operations that need a dup instruction are those binary operations whose both operands are away from QH. The augmented LDAG with ghost nodes facilitates the task of identifying those instructions. All binary operations having ghost nodes as their left and right children need to be transformed as follows.

The ghost node in the left children is substituted by a dup node, and the ghost node in the right children is eliminated from the LDAG. For those binary operations with only one ghost node as the left or right children, the ghost node is eliminated from the LDAG. Algorithm 7.2 describes the function dup_assignment().

Algorithm 7.2 dup_assignment (Node i)

1: **if** isBinary (i) **then**
2:    **if** isGhost (i.left) and isGhost (i.right) **then**
3:       dup_assign_node (i.left)
4:       dag_remove_node (i.right)
5:    **else if** isGhost (i.left) **then**
6:       dag_remove_node (i.left)
7:    **else if** isGhost (i.right) **then**
8:       dag_remove_node (i.right)
9:    **end if**
10:   **return**
11: **end if**

7.3.2 Offset Calculation Phase

Once the LDAGs including dup instructions have been built, the next step is to calculate the offset reference values for the instructions. Following the definition of the producer order QCM, the offset reference value of an instruction represents the distance, in number of queue words, between the position of QH and the operand to be dequeued.

The main challenge in the calculation of offset values is to determine the QH relative position with respect of every operation. We define the following properties to facilitate the description of the algorithm to find the position of QH with respect of any node in the LDAG.

Definition 7.3.4 An α-node is the first element of a level.

Definition 7.3.5 The QH position with respect of the α-node of Level-j is always at the α-node of the next level, Level-(j+1).

Definition 7.3.6 A level-order traversal of a LDAG is a walk of all nodes in every level (from the deepest to the root) starting from the α-node.

Definition 7.3.7 The distance between two nodes in a LDAG, $\delta(u, v)$, is the number of nodes found in a level-order traversal between u and v including u.

Definition 7.3.8 A hard edge is a dependence edge between two nodes that spans only one level.

Let p_n be a node for which the QH position must be found. QH relative position with respect of p_n is found after a node in a traversal P_i from p_{n-1} to p_0 (α-node) meets one of two conditions. The first condition is that the node is the α-node, $P_i = p_0$. From Definition 7.3.5, QH position is at α-node of the next level $lev(p) + 1$. The second condition is that P_i is a binary or unary operation and has a hard edge to one of its operands q_m. QH position is given by q_m's following node as a result of a level-order traversal. Notice that q_m's following node can be q_{m+1}, or the α-node of $lev(q_m) + 1$ if q_m is the last node in $lev(q_m)$. The proposed algorithm is described in Algorithm 7.3.

After the QH position with respect of p_n has been found, the only operation to calculate the offset reference value for each of p_n's operands is to measure the distance δ between QH's position and the operand's position as described in Algorithm 7.4.

In brief, for all nodes in a LDAG w, the offset reference values to their operands are calculated by determining the position of QH with respect of every node, and measuring the distance to the operands. Every edge is annotated with its offset reference value.

Algorithm 7.3 qh_pos (LDAG w, Node u)

1: $I \Leftarrow$ getLevel (u)
2: **for** $i \Leftarrow u.prev$ to $I.\alpha$-node **do**
3:      **if** isOperation (i) **then**
4:          **if** isHardEdge $(i.right)$ **then**
5:              $v \Leftarrow$ BFS_nextnode $(i.right)$
6:              **return** v
7:          **end if**
8:          **if** isHardEdge $(i.left)$ **then**
9:              $v \Leftarrow$ BFS_nextnode $(i.left)$
10:             **return** v
11:         **end if**
12:     **end if**
13: **end for**
14: $L \Leftarrow$ getNextLevel (u)
15: $v \Leftarrow L.\alpha$-node
16: **return** v

Algorithm 7.4 OpOffset (LDAG w, Node v, Operand r)

1: offset $\Leftarrow \delta$(qh_pos$(w, v), r$)
2: **return** offset

7.3.3 Instruction Scheduling Phase

The instruction scheduling algorithm of our compiler is a variation of basic block scheduling [1] where the only difference is that instructions are generated from a level-order topological order of the LDAGs. The input of the algorithm is an LDAG annotated with offset reference values. For every level in the LDAG, from the deepest level to the root level, all nodes are traversed from left to right and an equivalent low-level intermediate representation instruction is selected for every visited node.

Fig. 7.4 QIR code fragment

```
(QMARK_BBSTART (B1))

(QMARK_STMT)
    (QMARK_LEVEL)
        (PUSH_Q (i))
        (PUSH_Q (4))
    (QMARK_LEVEL)
        (LOAD_ADDR_Q (a))
        (MUL_Q)
    (QMARK_LEVEL)
        (ADD_Q)
    (QMARK_LEVEL)
        (SLOAD_Q)
    (QMARK_LEVEL)
        (POP_Q (x))

(QMARK_STMT)
    (QMARK_LEVEL)
        (PUSH_Q (x))
        (PUSH_Q (4))
    (QMARK_LEVEL)
        (LOAD_ADDR_Q (a))
        (MUL_Q)
    (QMARK_LEVEL)
        (ADD_Q)
        (PUSH_Q (7))
    (QMARK_LEVEL)
        (STORE_Q)

(QMARK_STMT)
    (GOTO_Q (L2))
```

Instruction selection was simplified by having one low-level instruction for every high-level instruction in the LDAG representation. The output of the instruction scheduling is a QIR list. QIR is a single-operand low-level intermediate representation capable to express the instruction set of the QueueCore. The only operand is used for memory operations and branch instructions. Offset reference values are encoded as attributes in the QIR instructions. Figure 7.4 shows the QIR list for the LDAG. The QIR includes annotations depicted in Fig. 7.4 with the prefix QMARK_*.

An extra responsibility of this phase is to check code correctness of the one-offset P-Code generation algorithm by comparing with zero the value of the offset reference for the first operand of binary instructions based on the assumption that the one-offset P-Code generation algorithm constrains all instructions to have at most one offset reference. For every compiled function, this phase also inserts the QIR instructions for the function's prologue and epilogue.

7.3.4 Natural Instruction Level Parallelism Extraction: Statement Merging Transformation

Statement merging transformation reorders the instructions of a sequential program in such a way that all independent instructions from different statements are in the same level and can be executed in parallel following the principle of the QCM. This phase makes a dependence analysis on individual instructions of different statements looking for conflicts in memory locations. Statements are considered the transformation unit. Whenever an instruction is reordered, the entire data flow graph of the statement to where it belongs is reordered to keep its original shape. In this way, all offsets computed by the offset calculation phase remain the same, and the data flow graph is not altered.

The data dependence analysis looks for two accesses to the same memory location whenever two instructions have the same offset with respect of the base register. Instructions that may alias memory locations are merged safely using a conservative approach to guarantee correctness of the program. Statements with branch instructions and function calls are non-mergeable.

Figure 7.5a shows a program with three statements S_1, S_2, S_3. The original sequential scheduling of this program is driven by a level-order scheduling as shown in Fig. 7.5b. When the statement merging transformation is applied to this program, a dependency analysis reveals a flow dependency for variable x in S_1, S_2 in levels L_4, L_3. Instructions from S_2 can be moved one level down and the flow dependency on variable x is kept as long the store to memory happens before the load. Statement S_3 is independent from the previous statements, this condition allows S_3 to be pushed to the bottom of the data flow graph. Figure 7.5c shows the DFG for the sample program after the statement merging transformation. For this example, the number of levels in the DFG has been reduced from seven to five.

Fig. 7.5 Statement merging transformation

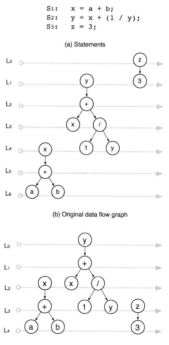

```
S1:    x = a + b;
S2:    y = x + (1 / y);
S3:    z = 3;
```

(a) Statements

(b) Original data flow graph

(c) Data flow graph after statement merging transformation

From the QCM principle, the QueueCore is able to execute the maximum parallelism found in DAGs as no false dependencies occur in the instructions. This transformation merges statements to expose all the available parallelism [35] within basic blocks. With the help of the compiler, QueueCore is able to execute `natural` instruction level parallelism as it appears in the programs. Statement merging is available in the queue compiler as an optimization flag which can be enabled upon user request.

7.3.5 Assembly Generation Phase

The last stage of the queue compiler is the assembly code generation for the QueueCore processor. It is done by a one-to-one translation from QIR code to assembly code. The assembly generator is in charge of inserting `covop` instructions to expand the operand field of those instructions that have operands beyond the limits of the operand field bits.

Figure 7.6a shows the generated assembly code and Fig. 7.6b shows the assembly code with natural parallelism exposed for the C program. Notice that the original assembly code and the assembly code after statement merging contain exactly the same instructions with the only difference that the order of the instructions change. All instructions have one operand. Depending on the instruction type, the only operand has different meanings. The highlighted code fragment in Fig. 7.6a shows the assignment of an array element indexed by variable to another variable, in C language "x=a[i]". The first instruction loads the index variable into the queue, its operand specifies the base register and the offset to obtain the memory location of the variable.

```
          ld    ($fp)0  # y                        ld    ($fp)0  # y
          ldil  1                                  ldil  1
          ceq   1       # compare equal            ceq   1       # compare equal
          bt    L1      # branch true              bt    L1      # branch true
    L0:                                       L0:
          ld    ($fp)8  # i                         ld    ($fp)8  # i
          ldil  4       # size of a[] element       ldil  4       # size of a[] element
          lda   ($fp)12 # load address of a[0]      lda   ($fp)12 # load address of a[0]
          mul   1       # i * size(a[])             mul   1       # i * size(a[])
          add   1       # address of a[i]           ld    ($fp)4  # x
          lds   0       # load computed address     ldil  4       # size of a[] element
          st    ($fp)4  # x                         add   1       # address of a[i]
          ld    ($fp)4  # x                         lda   ($fp)12 # load address of a[0]
          ldil  4       # size of a[] element       mul   1       # x * size(a[])
          lda   ($fp)12 # load address of a[0]      lds   0       # load computed address
          mul   1       # x * size(a[])             add   1       # address of a[x]
          add   1       # address of a[x]           ldil  7       # rhs constant
          ldil  7       # rhs constant              st    ($fp)4  # x
          sst   0       # st constant in            sst   0       # st constant in
                        # computed address                        # computed address
          j     L2                                  j     L2
    L1:                                       L1:
          ld    ($fp)4  # x                         ld    ($fp)4  # x
          ldil  2                                   ldil  2
          mul   1                                   mul   1
          covop 3       # (3<<8) = 768              covop 3       # (3<<8) = 768
          ldil  232     # 768 + 232 = 1000          ldil  232     # 768 + 232 = 1000
          add   1                                   add   1
          st    ($fp)4  # x                         st    ($fp)4  # x
    L2:                                       L2:

      (a) Original QueueCore assembly code        (b) ILP exposed QueueCore Assembly Code
```

Fig. 7.6 Assembly output for QueueCore processor

The operand in the second instruction specifies the immediate value to be loaded, if the value is greater than the instruction bits, the assembly phase inserts a `covop` instruction to extend the immediate value. The operand in the third instruction works is used to compute the effective address of the first element of the array. The next two arithmetic instructions use their operand as the offset reference and help to compute the address of the array element indexed by a variable. For this example, both are binary instructions and take their first operand implicitly from QH and the second operand from QH+1. The `lds` instruction loads into the queue the value of a computed address taken the operand queue as an offset reference given by its only operand. The last instruction stores the value pointed by QH to memory using base addressing.

To demonstrate the efficiency of our one-offset queue computation model, we developed a C compiler that targets the QueueCore processor. For a set of numerical benchmark programs, we evaluated the characteristics of the resulting queue compiler. We measured the effectiveness of statement merging optimization for improving ILP, we analyzed the quality of the generated code in terms of the distribution of instruction types, and we demonstrate the effectiveness of the queue compiler as a design space exploration tool for our QueueCore by analyzing the maximum offset value required by the chosen numerical benchmarks.

To show the potential of our technique for a high-performance processor, we compared the compile time exposed ILP from our compiler against the ILP exposed by an optimizing compiler for a typical RISC processor. And to highlight the low code size features of our design, we also compare the code size to the embedded versions of two RISC processors.

The chosen benchmarks are well-known numerical programs: radix-8 fast Fourier transform, livermore loops, whetstone loops, single precision linpack, and quake benchmark. To compare the extracted ILP, we compiled the programs using our queue compiler with statement merging transformation. For the RISC-like processor, we compiled the benchmarks using GCC 4.0.2 with classical and ILP optimizations enabled (-O3) targeting the MIPS I [9] instruction set. The ILP for the QueueCore is measured directly from the DDG in the compiler. The ILP for the MIPS I is measured from the generated assembly based on the register and memory data dependencies and control flow, assuming no-aliasing information.

Code size was measured from the text segment of the compiled programs. MIPS16 [36] and ARM/Thumb [37] were chosen for the RISC-like embedded processors. GCC 4.0.2 compiler for MIPS16 and ARM/Thumb architectures was used with full optimizations enabled (-O3) to generate the object files. For the QueueCore, the queue compiler was used with statement merging transformation.

7.4 Parallelizing Compiler Development Results

The resulting back end for the QueueCore consists of about 8000 lines of C code. Table 7.1 shows the number of lines for each phase of the back end.

Table 7.1 Lines of C code for each phase of the queue compiler's back end

Phase	Lines of code	Description
One-offset P-Code generation	3000	Code generation algorithm, QTrees and LDAGs infrastructure
Offset calculation	1500	Algorithm to find the location of QH and distance to each operation
Instruction scheduling	1500	Level-order scheduling, lowering to QIR, and QIR infrastructure
Statement merging	1000	Natural ILP exploitation and data dependency analysis
Assembly generation	1000	Assembly code generation from QIR
Total	**8000**	

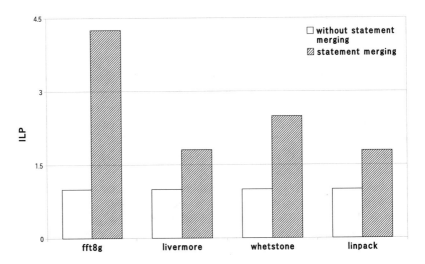

Fig. 7.7 Effect on ILP of statement merging transformation in the queue compiler

7.4.1   Queue Compiler Evaluation

First, we analyze the effect of the statement merging transformation on boosting ILP in our compiler. Figure 7.7 shows the improvement factor of the compiled code with statement merging transformation over the original code without statement merging, both scheduled using the level-order traversal.

All benchmarks show an improvement gain ranging from 1.73 to 4.25. The largest ILP improvement is for the fft8g program because it contains very large loop bodies without control flow where the statement merging transformation can work most effectively. Statement merging is a code motion transformation and does not insert or eliminate instructions.

Table 7.2 Instruction category percentages for the compiled benchmarks for the QueueCore

Benchmark	Memory	ALU	Move data	Ctrl. flow	Covop
fft8g	48.60	47.55	0.32	2.90	0.63
Livermore	58.55	33.29	0.20	5.95	4.01
Whetstone	58.73	26.73	1.11	13.43	0
Linpack	48.14	41.59	0.58	8.16	1.52
Equake	44.52	43.00	0.56	7.76	3.5

Table 7.3 QueueCore's program maximum offset reference value

Benchmark	Maximum offset
fft8g	29
Livermore	154
Whetstone	31
Linpack	174
Equake	89

To evaluate the quality of the generated code of our compiler, we organized the QueueCore instructions into five categories: memory, ALU, move data, control flow, and covop. Memory instructions are to load and store to main memory including loading immediate values; ALU includes comparison instructions, type conversions, integer and floating point arithmetic–logic instructions; move data includes all data transfer between special purpose registers; control flow includes conditional and unconditional jumps, and subroutine calls; and covop includes all covop instructions to extend memory accesses and immediate values.

Table 7.2 shows the distribution of the instruction categories in percentages for the compiled programs. From the table, we can observe that memory operations account for about 50% of the total number of instructions, ALU instructions about 40%, move data less than 1%, control flow less about 8%, and covop about 2%. These results point a place for future improvement of our compiler infrastructure. We believe that classical local and global optimizations [38] may improve the quality of the generated code by reducing the number of memory operations.

The developed queue compiler is a valuable tool for the QueueCore's architecture design space exploration since it gives us the ability to automatically generate assembly code and extract characteristics of the compiled programs that affect the processor's parameters. To emphasize the usage of the queue compiler as a design tool, we measured the maximum offset value required by the compiled benchmarks.

Table 7.3 shows the maximum offset value for the given programs. These compiling results show that the eight bits reserved in the QueueCore's instruction format [39] for the offset reference value are enough to satisfy the demands of these numerical calculation programs.

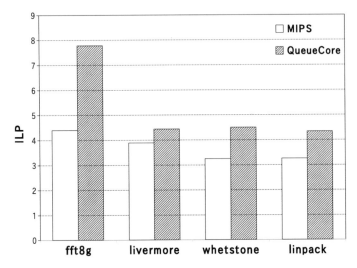

Fig. 7.8 Instruction level parallelism improvement of queue compiler over optimizing compiler for a RISC machine

7.4.2 Comparison of Generated QueueCore Code with Optimized RISC Code

The graph in Fig. 7.8 compares the ILP improvement of the queue compiler over the optimizing compiler for MIPS processor. For all the analyzed programs, the queue compiler exposed more natural parallelism to the QueueCore than the optimizing compiler for the RISC machine. The improvement of parallelism comes from the natural parallelism found in the level-order scheduled data flow graph with merged statements.

QueueCore's instruction set benefits from this transformation and scheduling as no register names are present in the instruction format. The RISC code, on the other hand, is limited by the architected registers. It depends on the good judgment of the compiler to make effective use of the registers to extract as much parallelism as possible, and whenever the register pressure exceeds the limit then spill registers to memory. The loop bodies in livermore, whetstone, linpack, and equake benchmarks consist of one or few instructions with many operands and operations. The improvement of our technique in these programs comes mainly from the level-order scheduling of these "fat" statements since the statement merging has no effect across basic blocks.

The greatest improvement on these benchmarks was for the fft8g program which is dominated by manually unrolled loop bodies where the statement merging takes full advantage. In average, our queue compiler is able to extract more parallelism than the optimizing compiler for a RISC machine by a factor of 1.38.

Figure 7.9 shows the normalized code size of the compiled benchmarks for MIPS16, ARM/Thumb and QueueCore using the MIPS I as the baseline. For most of

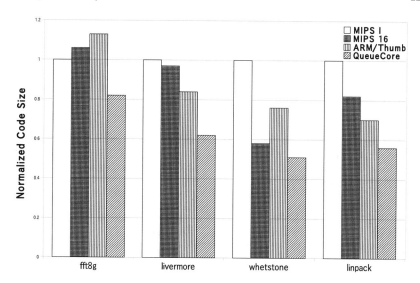

Fig. 7.9 Normalized code size for two embedded RISC processors and QueueCore

the benchmarks, our design achieves denser code than the baseline and the embedded RISC processors. Except for the equake program, where the MIPS16 achieved lower code size than the QueueCore.

A closer inspection of the object file revealed that the QueueCore program has about two times more instructions than the MIPS and MIPS16 code. This is due to the effect of local optimizations such as constant folding, common sub expression elimination, dead code removal, etc., that are applied in the RISC compilers and not in the queue compiler. On average, our design achieves 31% denser code than MIPS I, 20% denser code than the embedded MIPS16, and 26% denser code than the ARM/Thumb processor.

7.5 Chapter Summary

With the rise of multicore systems and many-core processors, concurrency becomes a major issue in the daily life of a programmer. Thus, compiler and software development tools will be critical to help programmers create high-performance software.

This chapter described design and evaluation of a parallelizing compiler targeted for single- and multicore computing. The design eliminates the register pressure by hiding completely the register file from the instruction set while maintaining a short instruction format with one operand reference. The queue compiler takes advantage of this design and it is capable to expose the maximum natural parallelism available in the data flow graph by means of a statement merging transformation. We evaluated the design by comparing the compile time extracted parallelism against an optimizing compiler for a traditional RISC machine for a set of numerical benchmarks.

References

1. S.S. Muchnick, *Advanced Compiler Design and Implementation* (Morgan Kaufman, Burlington, 1997)
2. J. Hennessy, D. Patterson, *Computer Architecture: A Quantitative Approach* (Morgan Kaufman, Burlington, 1990)
3. R. Allen, K. Kennedy, *Optimizing Compilers for Modern Architectures*, (Morgan Kaufman, Burlington, 2002)
4. M. Wolfe, *High Performance Compilers for Parallel Computing* (Addison-Wesley, 1996)
5. M. Lam, Software pipelining: an effective scheduling technique for VLIW machines, in *Proceedings of the ACM SIGPLAN 1988 conference on Programming Language design and Implementation*, (1988), pp. 318–328
6. R. Rau, Iterative modulo scheduling: an algorithm for software pipelining loops, in *Proceedings of the 27th annual international symposium on Microarchitecture*, (1994), pp. 63–74
7. J. Losa, E. Ayguade, M. Valero, Quantitative evaluation of register pressure on software pipelined loops. Int. J. Parallel Program. **26**(2), 121–142 (1998)
8. S. Pinter, Register allocation with instruction scheduling, in *Proceedings of the ACM SIGPLAN 1993 Conference on Programming Language Design and Implementation*, (1993), pp. 248–257
9. G. Kane, J. Heinrich, *MIPS RISC Architecture*, (Prentice Hall, 1992)
10. R. Kessler, The Alpha 21264 microprocessor. IEEE Micro **19**(2), 24–36 (1999)
11. Sparc-International, *The SPARC Architecture Manual, Version 8*, (Prentice Hall, 1992)
12. S.A. Mahlke, W.Y. Chen, P.P. Chang, W. mei, W. Hwu, Scalar program performance on muliple-instruction-issue processors with a limited number of registers, in *Proceedings of the 25th Annual Hawaii Int'l Conference on System Sciences*, (1992), pp. 34–44
13. M. Postiff, D. Greene, T. Mudge, *The Need for Large Register File in Integer Codes, Technical Report CSE-TR-434-00*, (University of Michigan, 2000)
14. J. Janssen, H. Corporaal, Partitioned register file for TTAs, in *Proceedings of the 28th annual international symposium on Microarchitecture*, (1995), pp. 303–312
15. J. Zalamea, J. Llosa, E. Ayguade, M. Valero, Software and hardware techniques to optimize register file utilization in VLIW architectures. Int. J. Parallel Program. **32**(6), 447–474 (2004)
16. X. Huang, S. Carr, P. Sweany, Loop transformations for architectures with partitioned register banks, in *Proceedings of the ACM SIGPLAN Workshop on Languages, Compilers and Tools for Embedded Systems*, (2001), pp. 48–55
17. S. Jang, S. Carr, P. Sweany, D. Kuras, A code generation framework for VLIW architectures with partitioned register banks, in *Proceedings of the 3rd International Conference on Massively Parallel Computing Systems*, (1998)
18. M. Fernandes, J. Llosa, N. Topham, *Using Queues for Register File Organization in VLIW. Technical Report ECS-CSG-29-97*, (University of Edinburgh, Department of Computer Science, 1997)
19. G. Tyson, M. Smelyanskiy, E. Davidson, Evaluating the use of register queues in software pipelined loops. IEEE Trans. Comput. **50**(8), 769–783 (2001)
20. R. Ravindran, R. Senger, E. Marsman, G. Dasika, M. Guthaus, S. Mahlke, R. Brown, Partitioning variables across register windows to reduce spill code in a low-power processor. IEEE Trans. Comput. **54**(8), 998–1012 (2005)
21. G. Kucuk, O. Ergin, D. Ponomarev, K. Ghose, Energy efficient register renaming. Lect. Notes Comput. Sci. **2799**(2003), 219–228 (2003)
22. B. Preiss, C. Hamacher, Data flow on queue machines, in *12th International IEEE Symposium on Computer, Architecture*, (1985), pp. 342–351
23. S. Okamoto, Design of a superscalar processor based on queue machine computation model, in *IEEE Pacific Rim Conference on Communications, Computers and Signal Processing*, (1999), pp. 151–154
24. H. Schmit, B. Levine, B. Ylvisaker, Queue machines: hardware compilation in hardware, in *FCCM'02, 10th Annual IEEE Symposium on Field-Programmable Custom Computing Machines*, (2002), pp. 152–161

25. A. Ben Abdallah, T. Yoshinaga, M. Sowa, High-level modeling and FPGA prototyping of produced order parallel queue processor core. J. Supercomput. **38**(1), 3–15 (2006)
26. M. Sowa, A. Ben Abdallah, T. Yoshinaga, Parallel queue processor architecture based on produced order computation model. J. Supercomput. **32**(3), 217–229 (2005)
27. A. Canedo, A. Ben Abdallah, M. Sowa, A GCC-based Compiler for the queue register processor, in *Proceedings of International Workshop on Modern Science and Technology*, (May 2006), pp. 250–255
28. A. Ben Abdallah, M. Masuda, A. Canedo, K. Kuroda, Natural instruction level parallelism-aware compiler for high-performance queuecore processor architecture. J. Supercomput. **57**(3), 314–338 (2011)
29. A. Ben Abdallah, A. Canedo, T. Yoshinaga, M. Sowa, The QC-2 parallel queue processor architecture. J. Parallel Distrib. Comput. **68**(2), 235–245 (2008)
30. A. Canedo, A. Ben Abdallah, M. Sowa, A new code generation algorithm for 2-offset producer order queue computation model. J. Comput. Lang. Syst. Struct. **34**(4), 184–194 (2007)
31. A. Canedo, A. Ben Abdallah, M. Sowa, Compiling for reduced bit-width queue processors. J. Signal Process. Syst. **59**(1), 45–55 (2010)
32. A. Canedo, A. Ben Abdallah, M. Sowa, Efficient compilation for queue size-constrained queue processors. J. Parallel Comput. **35**, 213–225 (2009)
33. D. Novillo, Design and implementation of tree SSA, in *Proceedings of GCC Developers Summit*, (2004), pp. 119–130
34. L.S. Heath, S.V. Pemmaraju, Stack and queue layouts of directed acyclic graphs: part I. SIAM J. Comput. **28**(4), 1510–1539 (1999)
35. D. Wall, Limits of instruction-level parallelism. ACM SIGARCH Comput. Archit. News **19**(2), 176–188 (1991)
36. K. Kissel, *MIPS16: High-density MIPS for the Embedded Market* (Technical report, Silicon Graphics MIPS Group, 1997)
37. L. Goudge, S. Segars, Thumb: reducing the cost of 32-bit RISC performance in portable and consumer applications, in *Proceedings of COMPCON 1996*, (1996), pp. 176–181
38. A.V. Aho, R. Sethi, J.D. Ullman, *Compilers Principles, Techniques, and Tools*, (Addison Wesley, 1986)
39. A. Ben Abdallah, S. Kawata, M. Sowa, Design and architecture for an embedded 32-bit queuecore. J. Embed. Comput. **2**(2), 191–205 (2006)

Chapter 8
Power Optimization Techniques for Multicore SoCs

Abstract Power dissipation continues to be a primary design constraint in single and multicore systems. Increasing power consumption not only results in increasing energy costs, but also results in high die temperatures that affect chip reliability, performance, and packaging cost. Energy conservation has been largely considered in the hardware design in general and also in embedded multicore systems' components, such as CPUs, disks, displays, memories, and so on. Significant additional power savings can be also achieved by incorporating low-power methods into the design of network protocols used for data communication (audio, video, etc.). This chapter investigates in details power reduction techniques at components and the network protocol levels.

8.1 Introduction

Computation and communication have been steadily moving toward embedded multicore devices. With continued miniaturization and increasing computation power, we see ever growing use of powerful microprocessors running sophisticated, intelligent control software in a vast array of devices including pagers, cellular phones, laptop computers, digital cameras, video cameras, video games, etc. Unfortunately, there is an inherent conflict in the design goals behind these devices: as mobile systems, they should be designed to maximize battery life, but as intelligent devices, they need powerful processors, which consume more energy than those in simpler devices, thus reducing battery life.

In spite of continuous advances in semiconductor and battery technologies that allow microprocessors to provide much greater computation per unit of energy and longer total battery life, the fundamental trade-offs between performance and battery life remain critically important [1–4].

Multimedia applications and mobile computing are two trends that have a new application domain and market. Personal mobile or ubiquitous computing is playing a significant role in driving technology. An important issue for these devices will be the user interface—the interaction with its owner. The device needs to support multimedia tasks and handles many different classes of data traffic over a limited bandwidth wireless connection, including delay-sensitive, real-time traffic such as video and speech.

Wireless networking greatly enhances the utility of a personal computing device. It provides mobile users with versatile communication and permits continuous access to services and resources of the land-based network. A wireless infrastructure capable of supporting packet data and multimedia services in addition to voice will boot-strap on the success of the Internet, and in turn drives novel networked applications and services. However, the technological challenges to establishing this paradigm of personal mobile computing are nontrivial. In particular, these devices have lim-ited battery resources. While reduction of the physical dimensions of batteries is a promising solution, such effort alone will reduce the amount of charge retained by the batteries. This will in turn reduce the amount of time a user can use the comput-ing device. Such restrictions tend to undermine the notion of mobile computing. In addition, more extensive and continuous use of network services will only aggravate this problem since communication consumes relatively much energy. Unfortunately, the rate at which battery performance improves is very slow, despite the great interest created by the wireless business.

The energy efficiency is an issue involving all layers of the system, its physical layer, its communication protocol stack, its system architecture, its operating system, and the entire network [1]. This implicates several mechanisms that can be used to attain a high-energy efficiency. There are several motivations for energy-efficient design. Perhaps, the most visible driving source is the success and growth of the portable consumer electronic market.

In its most abstract form, a networked system has two sources of energy drain required for its operation:

1. Communication, due to energy spent by the wireless interface and due to the internal traffic between various parts of the system, and
2. Computation, due to processing for applications, tasks required during commu-nication, and operating system.

Thus, minimizing energy consumption is a task that will require minimizing the contributions of communication and computation.

From another hand, power consumption has become a major concern because of the ever-increasing density of solid-state electronic devices, coupled with an increas-ing use of mobile computers and portable communication devices. The technology has thus far helped to build low-power systems. The speed–power efficiency has indeed gone up since 1990 by 10 times each 2.5 years for general-purpose proces-sors and digital signal processors (DSPs) [4].

Design for low-energy consumption is certainly not a challenging research field, and yet remains one of the most difficult as future mobile multicore SoC system designers attempt to pack more capabilities such as multimedia processing and high bandwidth radios into battery-operated portable miniature packages. Playing times of only a few hours for personal audio, notebooks, and cordless phones are clearly not very consumer friendly. Also, the required batteries are voluminous and heavy, often leading to bulky and unappealing products [5].

The key to energy efficiency in future mobile multicore SoCS will be, then, designing higher layers of the mobile system, their functionality, their system architecture, their operating system, and the entire network, with energy efficiency in mind.

8.2 Power-Aware Technological-Level Design Optimizations

8.2.1 Factors Affecting CMOS Power Consumption

Most components in a mobile system are currently fabricated using CMOS technology. Since CMOS circuits do not dissipate power if they are not switching, a major focus of low-power design is to reduce the switching activity to the minimal level required to perform the computations [6, 7].

The sources of energy consumption on a CMOS chip can be classified as static and dynamic power dissipation. The average power is given by

$$P_{avg} = P_{static} + P_{dynamic}. \tag{8.1}$$

The static power consumption is given by

$$P_{static} = P_{short\text{-}circuit} + P_{leak} = I_{sc} \cdot V_{dd} + I_{leak} \cdot V_{dd} \tag{8.2}$$

and the dynamic power consumption is given by

$$P_{dynamic} =_{\alpha_0 \to 1} C_L \cdot V_{dd}^2 \cdot f_{clk}. \tag{8.3}$$

The three major sources of power dissipation are, then, summarized in the following equation:

$$P_{avg} =_{\alpha_0 \to 1} C_L \cdot V_{dd}^2 \cdot f_{clk} + I_{sc} \cdot V_{dd} + I_{leak} \cdot V_{dd}. \tag{8.4}$$

The first term of formula 4 represents the switching component of power, where $\alpha_0 \to 1$ is the node transition activity factor (the average number of times the node makes a power-consuming transition in one clock period), C_L is the load capacitance, and f_{clk} is the clock frequency. The second term is due to the direct-path short-circuit current, I_{sc}, which arises when both the NMOS and PMOS transistors are simultaneously active, conducting current directly from supply ground. The last term, I_{leak} (leakage current), which can arise from substrate injection and sub-threshold effects, is primarily determined by fabrication technology.

$\alpha_0 \to 1$ is defined as the average number of times in each clock cycle that a node with capacitance, C_L, will make a power-consuming transition resulting in an average switching component of power for a CMOS gate to be simplified to

$$P_{\text{switch}} =_{\alpha_0 \to 1} C_L \cdot V_{\text{dd}}^2 \cdot f_{\text{clk}}. \tag{8.5}$$

Since the energy expended for each switching event in CMOS circuits is $C_L \cdot V_{\text{dd}}^2 \cdot f_{\text{clk}}$, it has the extremely important characteristics that it becomes quadratically more efficient as the high transition voltage level is reduced.

It is clear that operating at the lowest possible voltage is most desirable; however, this comes at the cost of increased delays and thus reduced throughput. It is also possible to reduce the power by choosing an architecture that minimizes the effective switched capacitance at a fixed voltage: through reductions in the number of operations, the interconnect capacitance, internal bit widths, and using operations that require less energy per computation. We will use Formulas (8.4) and (8.5) to discuss the energy reduction techniques and trade-offs that involve energy consumption of digital circuits. From these formulas, we can see that there are four ways to reduce power:

1. reduce the capacity load C,
2. reduce the supply voltage V,
3. reduce the switching frequency f, and
4. reduce the switching activity.

8.2.2   Reducing Voltage and Frequency

Supply voltage scaling has been the most adopted approach to power optimization, since it normally yields considerable savings, thanks to the quadratic dependence of P_{switch} on V_{dd} [6]. The major shortcoming of this solution, however, is that lowering the supply voltage affects circuit speed. As a consequence, both design and technological solutions must be applied in order to compensate the decrease in circuit performance introduced by reduced voltage. In other words, speed optimization is applied first, followed by supply voltage scaling, which brings the design back to its original timing, but with a lower power requirement.

It is well known that reducing clock frequency f alone does not reduce energy, since to do the same work the system must run longer. As the voltage is reduced, the delay increases. A common approach to power reduction is to first increase the speed performance of the module itself, followed by supply voltage scaling, which brings the design back to its original timing, but with lower power requirements [7].

A similar problem, i.e., performance decrease, is encountered when power optimization is obtained through frequency scaling. Techniques that rely on reductions of the clock frequency to lower power consumption are thus usable under the constraint that some performance slack does exist. Although this may seldom occur for designs considered in their entirety, it happens quite often that some specific units in a larger architecture do not need peak performance for some clock/machine cycles. Selective frequency scaling (as well as voltage scaling) on such units may thus be applied, at no penalty in the overall system speed.

8.2.3 Reducing Capacitance

Energy consumption in CMOS circuitry is proportional to capacitance C. There-fore, a path that can be followed to reduce energy consumption is to minimize the capacitance. A significant fraction of a CMOS chips energy consumption is often contributed to driving large off-chip capacitances, and not to core processing. Off-chip capacitances are in the order of five to tens of pF. For conventional packaging technologies, pins contribute approximately 13–14 pF of capacitance each (10 pF for the pad and 3–4 pF for the printed circuit board) [8].

From our earlier discussion, Eq. (8.5) indicates that energy consumption is pro-portional to capacitance; I/O power can be a significant portion of the overall energy consumption of the chip. Therefore, in order to save energy, use few external outputs, and have them switch as infrequently as possible. Packaging technology can have a impact on the energy consumption. For example, in multi-chip modules where all of the chips of a system are mounted on a single substrate and placed in a single pack-age, the capacitance is reduced. Also, accessing external memory consumes much energy. So, a way to reduce capacitance is to reduce external accesses and optimize the system by using on-chip resources like caches and registers.

8.2.3.1 Chip Layout

There are a number of layout-level techniques that can be applied. Since the physical capacitance of the higher metal layers is smaller, there is some advantage to select upper level metals to route high-activity signals. Furthermore, traditional placement involves reducing area and delay, which in turn translates to minimizing the physical capacitance of wires. Placement that incorporates energy consumption concentrates on minimizing the activity-capacitance product rather than capacitance alone. In general, high-activity wires should be kept short and local. Tools have been developed that use this basic strategy to achieve about 18% reduction in energy consumption.

The capacitance is an important factor for the energy consumption of a sys-tem. However, reducing the capacity is not the distinctive feature of low-power design, since in CMOS technology energy is consumed only when the capacitance is switched. It is more important to concentrate on the switching activity and the num-ber of signals that need to be switched. Architectural design decisions have more impact than solely reducing the capacitance.

8.2.3.2 Technology Scaling

Scaling advanced CMOS technology to the next generation improves performance, increases transistor density, and reduces power consumption. Technology scaling typically has three main goals:

1. Reduce gate delay by 30%, resulting in an increase in operating frequency of about 43%;
2. Double transistor density; and
3. Reduce energy per transistor by about 65%, saving 50% of the power.

These are not ad hoc goals; rather, they follow scaling theory [8].

As the Semiconductor Industry Association road map (SIA) indicates, the trend of process technology improvement is expected to continue for years [9]. Scaling of the physical dimension involves reducing all dimensions: thus transistors widths and lengths are reduced; interconnection length is reduced, etc. Consequently, the delay, capacitance, and energy consumption will decrease substantially.

Another way to reduce capacitance at the technology level is to reduce chip area. For example, an energy-efficient architecture that occupies a larger area can reduce the overall energy consumption, e.g., by exploiting locality in a parallel implementation.

8.3 Power-Aware Logic-Level Design Optimizations

Logic-level power optimization has been extensively researched in the last few years. While most traditional power optimization techniques for logic cells focus on minimizing switching power, circuit design for leakage power reduction is also gaining importance [10]. As a result, logic-level design can have a high impact on the energy efficiency and performance of the system. Issues in the logic level relate to, for example, state machines, clock gating, encoding, and the use of parallel architectures.

8.3.1 Clock Gating

Several power minimization techniques work especially well at the logic level. Most of them rely on switching frequency. The best example of which is the use of clock gating [11]. Clock gating provides a way to selectively stop the clock, and thus force the original circuit to make no transition, whenever the computation to be carried out by a hardware unit at the next clock cycle is useless. In other words, the clock signal is disabled to shut down some modules of the chip that are inactive. This saves on clock power, because the local clock line is not toggling all the time.

For example, the latency for the CPU of the TMS320C5x DSP processor [12] to return to active operation from the IDLE3 mode takes around $50\,\mu s$, due to the need of the on-chip PLL circuit to lock with the external clock generator. With the conventional scheme, the register is clocked all the time, whether new data is to be captured or not. If the register must hold the old state, its output is fed back into the data input through a multiplexer whose enable line (ENABLE) controls whether the register clocks in new data or recycles the existing data. However, with a gated

Fig. 8.1 Clock gating
example

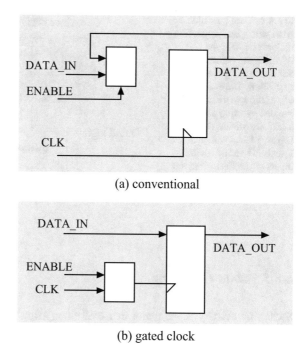

(a) conventional

(b) gated clock

clock, the signal that would otherwise control the select line on the multiplexer now controls the gate. The result is that the energy consumed in driving the register's clock input (CLK) is reduced in proportion to the decrease in average local clock frequency. The two circuits function identically, but utilization of the gated clock reduces the power consumption (Fig. 8.1).

8.3.2   Logic Encoding

The power consumption can be also reduced by carefully minimizing the number of transitions. The designer of a digital circuit often has the freedom of choosing the encoding scheme. Different encoding implementations often lead to different areas, powers, and delay trade-offs. An appropriate choice of the representation of the signals can have a big impact on the switching activity.

The frequency of consecutive patterns in the traffic streams is the basis for the effectiveness of encoding mechanisms. For example, a program counter in a processor generally uses a binary code. On average two bits are changed for each state transition [13]. Using a Gray-code (single bit changes) can give interesting energy savings. However, a Gray-code incremental requires more transistors to implement than a ripple carry incrementer [13]. Therefore, a combination can be used in which only the most frequently changing LSB bits use a Gray code.

Fig. 8.2 Dual operation ALU with guard logic. The multiplexer does the selection only after both units have completed their evaluation. The evaluation of one of the two units is avoided by using a guard logic; two latches (L1 and L2) are placed with enable signals (s1 and s2) at the inputs of the shifter and the adder respectively

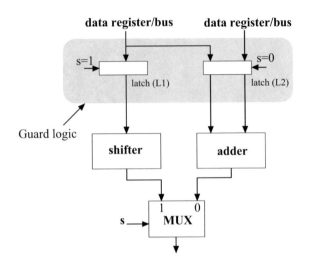

8.3.3   Data Guarding

Switching activity is the major cause of energy dissipation in most CMOS digital systems. Therefore, to reduce power consumption, switching activities that do not contribute to the actual communication and computation should be eliminated. The basic idea is to identify logical conditions at some inputs to a logic circuit that is invariant to the output. Since those input values do not affect the output, the input transitions can be disabled.

Data logic-guarding technique [14] is an efficient method used to guard not useful switching activities to propagate further inside the system. The technique is based on reducing the switching activities by placing transparent latches/registers with an enable signal at the inputs of each block of the circuit that needs to be selectively turned off. If the module is to be active in a clock cycle, the enable signal makes the latch transparent, permitting normal operation. If not, the latch retains its previous state and no transitions propagate through the inactive module (see Fig. 8.2). As a summary, the logic-level design can have a high impact on the energy efficiency and the performance of a given system. Even with the use of state-of-the-art hardware design language (i.e., Verilog HDL), there are still many optimizations techniques that should be explored by the designers to reduce the energy consumption at the logic level. The most effective technique used at this level is the reduction of switching activities.

8.4   Power-Aware System Level Design Optimizations

In the previous sections we have explored sources of energy consumption and showed the low level—technology and circuit levels, and design techniques used to reduce the power dissipation. In this section, we will concentrate on the energy reduction techniques at the architecture and system level.

8.4.1 Hardware System Architecture Power Consumption Optimizations

The implementation-dependent part of the power consumption of a system is strongly related to the number of properties that a given system or algorithm may have. The component that contributes a significant amount of the total energy consumption is the communication channels or interconnects.

Experiments have already been made in designs and proved that about 10–40% of the total power may be dissipated in buses, multiplexers and drivers [15, 16]. This amount can increase dramatically for systems with multiple chips due to large off-chip bus capacitance.

The energy consumption of the communication channels is largely dependent on algorithm and architecture-level design decisions. Regularity and locality are two important properties of algorithms and architectures for reducing the energy consumption due to the communication channels. The idea behind regularity is to capture the degree to which common patterns appear in an algorithm. Common patterns enable the design of less complex architecture and therefore simpler interconnect structure and less control hardware. Simple measures of regularity include the number of loops in the algorithm and the ratio of operations to nodes in the data flow graph. The statistics of the percentage of operations covered by sets of patterns is also indicative of an algorithm's regularity. Quantifying this measure involves first finding a promising set of patterns, large patterns being favored. The core idea is to grow pairs of as large as possible isomorphic regions from corresponding pairs of seed nodes [17].

Locality relates to the degree to which a system or algorithm has natural isolated clusters of operation or storage with few interconnections between them. Partitioning the system or algorithm into spatially local clusters ensures that the majority of the data transfers take place within the clusters and relatively few between clusters. The result is that the local buses with a low electrical capacity are shorter and more frequently used than the longer highly capacitive global buses. Locality of reference can be used to partition memories. Current high-level synthesis tools are targeted to area minimization or performance optimization. However, for power reduction it is better to reduce the number of accesses to long global buses and have the local buses be accessed more frequently.

In a direct implementation targeted at area optimization, hardware sharing between operations might occur, destroying the locality of computation. An architecture and implementation should preserve the locality and partition and implement it such that hardware sharing is limited. The increase in the number of functional units does not necessarily translate into a corresponding increase in the overall area and energy consumption since the localization of interconnect allows a more compact layout and also fewer access to buffers, and multiplexers are needed.

8.4.1.1 Hierarchical Memory System

Efficient use of an optimized custom memory hierarchy to exploit temporal locality in the data accesses can have a very large impact on the power consumption in data-dominated applications. The idea of using a custom memory hierarchy to minimize the power consumption is based on the fact that memory power consumption depends primarily on the access frequency and the size of the memory. For on-chip memories, memory power increases with the memory size. In practice, the relation is between linear and logarithmic depending on the memory library. For off-chip memories, the power is much less dependent on the size because they are internally heavily partitioned. Still they consume more energy per access than the smaller on-chip memories. Hence, power savings can be obtained by accessing heavily used data from smaller memories instead of from large background memories [18].

As most of the time only a small memory is read, the energy consumption is reduced. Memory considerations must also be taken into account in the design of any system. By employing an on-chip cache significant power reductions together with a performance increase can be gained. Apart from caching data and instructions at the hardware level, caching is also applied in the file system of an operating system [18]. The larger the cache is, the better performance is achieved. Energy consumption is reduced because data is kept locally, and thus requires less data traffic. Furthermore, the energy consumption is reduced because less disk and network activity is required.

The compiler also has impact on power consumption by reducing the number of instructions with memory operands. It also can generate code that exploits the characteristics of the machine and avoids expensive stalls. The most energy can be saved by a proper utilization of registers. In [19], a detailed review of some compiler techniques that are of interest in the power minimization arena is also presented.

Secondary Storage

Secondary storage in modern mobile systems generally consists of a magnetic disk supplemented by a small amount of DRAM used as a disk cache; this cache may be in the CPU main memory, the disk controller, or both [20–22]. Such a cache improves the overall performance of secondary storage. It also reduces its power consumption by reducing the load on the hard disk, which consumes more power than the DRAM.

Energy consumption is reduced because data is kept locally, and thus requires less data traffic. In addition, the energy consumption is reduced because less disk and network traffic is required. Unfortunately, there is trade-off in size of the cache memory since the required amount of additional DRAM can use as much as energy as a conventional spinning hard disk [23].

A possible technology for secondary storage is an integrated circuit called flash memory [21]. Like a hard disk, such memory is nonvolatile and can hold data without consuming energy. Furthermore, when reading or writing, it consumes only 0.15–0.47 W, far less than a hard disk. It has a read speed of about 85 ns per byte, quite like DRAM, but writes speed of about 410 μs per byte, about 10–100 times slower than hard disk. However, since flash memory has no seek time, its overall write performance is not that much worse than a magnetic disk; in fact, for sufficiently small random writes, it can actually be faster. Since flash is practically as fast as DRAM at reads, a disk cache in no longer important for read operation. The cost per megabyte of flash is about 7–40 times more expensive than guard disk, but about 2–5 times less expensive than DRAM. Thus, flash memory might also be effective as a second-level cache below the standard DRAM disk cache [20, 21].

8.4.1.2 Processor

In general, the power consumption of the CPU is related to the clock rate, the supply voltage, and the capacitance of the devices being switched [11, 24–26]. One power-saving feature is the ability to slow down the clock. Another is the ability to selectively shut off functional units, such as the floating-point unit; this ability is generally not externally controllable. Such a unit is usually turned off by stopping the clock propagated to it. Finally, there is the ability to shut down processor operation altogether so that it consumes little or no energy. When this last ability is used, the processor typically returns to full power when the next interrupt occurs. A time energy consumption relationship is shown in Fig. 8.3.

Turning off a processor has a little downside; no excess energy is expended turning the processor back on, the time until it comes back on is barely noticeable, and the state of the processor is unchanged from it turning off and on, unless it has a volatile cache [11]. Therefore, reducing the power consumption of the processor can have a greater effect on overall power savings than it might seem from merely examining the percentage of total power attributable to the processor.

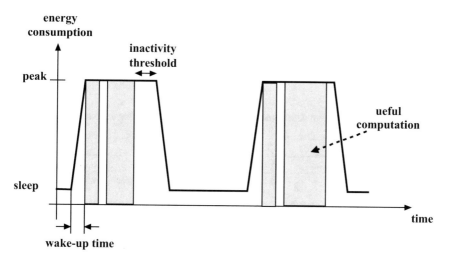

Fig. 8.3 Power consumption in typical processor core

8.4.1.3 Display and Backlight

The display and backlight have very few energy-saving features. This is unfortunate, since they consume a great deal of power in their maximum-power states; for instance, on the Duo 280c, the display consumes a maximum of 0.75 W and the backlight consumes a maximum of 3.40 [27, 28]. The backlight can have its power reduced by reducing the brightness level or by turning it off, since its power consumption is roughly proportional to the luminance delivered. The display power consumption can be reduced by turning the display off. It can also be reduced slightly by switching from color to monochrome or by reducing the update frequency, which reduces the range of shades or colors of Gray for each pixel, since such shading is done by electrically selecting each pixel for a particular fraction of its duty cycle. Generally, the only disadvantage of these low-power modes is reduced readability. However, in the case of switches among update frequencies and switches between color and monochrome, the transitions can also cause annoying flashes.

8.4.2 Operating System Power Consumption Optimization

Software and algorithmic considerations can also have a severe impact on energy consumption [2, 16, 19, 28–30]. Digital hardware designers have promptly reacted to the challenge posed by low-power design. Designer skills, technology improvements, and CAD tools have been successful in reducing the energy consumption. Unfortunately, software engineers and system architects are often less "energy-aware" than digital designers, and they also lack suitable tools to estimate the energy consumption of their designs. As a result, energy-efficient hardware is often employed in a way that does not make optimal use of energy-saving possibilities. In this section we will show several approaches to reduce energy consumption at the operating system level and to the applications (Table 8.1).

Table 8.1 Operating system functionality and corresponding techniques for optimizing energy utilization

CPU scheduling	Idle power mode, voltage scaling
Operating system functionality	Energy-efficient techniques
Memory allocation	Adaptive placement of memory blocks, switching of hardware energy reduction modes
Application/OS interaction	Agile content negotiation trading fidelity for power, APIs
Resource Protection and allocation	Fair distribution of battery life among both local and distributed tasks, locking battery for expensive operations
Communication	Adaptive network polling, energy-aware routing, placement of distributed computation, and server binding

A fundamental OS task is efficient management of host resources. With energy as the focus, the question becomes how to make the basic interactions of hardware and software as energy efficient as possible for local computation. One issue observed in traditional performance-centric resource management involves latency hiding techniques. A significant difference and challenge in energy-centric resource management is that power consumption is not easy to hide.

As one instance of power-aware resource management, we consider memory management. Memory instructions are among the more power-hungry operations on embedded processors [29], making the hardware/software of memory management a good candidate for optimization. Intels guidelines for mobile power [31, 32] indicate that the target for main memory should be approximately 4% of the power budget. This percentage can dramatically increase in systems with low-power processors, displays, or without hard disks. Since many small devices have no secondary storage and rely on memory to retain data, there are power costs for memory even in otherwise idle systems. The amount of memory available in mobile devices is expanding with each new model to support more demanding applications (i.e., multimedia), while the demand for longer battery life also continues to grow significantly.

Scheduling is needed in a system when multiple functional units need to access the same object. In operating systems scheduling is applied at several parts of a system for processor time, communication, disk access, etc. Currently, scheduling is performed on criteria like priority, latency, time requirements, etc. Power consumption is in general only a minor criterion for scheduling, despite the fact that much energy could be saved.

Subsystems of a computer, such as the CPU, the communication device, and storage system, have small usage duty cycles. That is, they are often idle and wait for the user or network interaction. Furthermore, they have huge differences in energy consumption between their operating states.

Recent advances in ad hoc networks allow mobile devices to communicate with one another, even in the absence of pre-existing base stations or routers. All mobile devices are able to act as routers, forwarding packets among devices that may otherwise be out of communication range of one another. Important challenges include discovering and evaluating available routes among mobile devices and maintaining these routes as devices move, continuously changing the "topology" of the underlying wireless network. In applications with limited battery power, it is important to minimize energy consumption in supporting this ad hoc communication.

There are numerous opportunities for power optimizations in such environments, including

(i) reducing transmission power adaptively based on the distance between sender and receiver,

(ii) adaptively setting transmission power in route discovery protocols,

(iii) balancing hop count and latency against power consumption in choosing the "best" route between two hosts, and

(iv) choosing routes to fairly distribute the routing duties (and the associated power consumption) among nodes in an ad hoc network [33].

8.4.3 Application, Compilation Techniques, and Algorithm

In traditional power-managed systems, the hardware attempts to provide automatic power management in a way that is transparent to the applications and users. This has resulted in some legendary user problems such as screens going blank during video or slide show presentations, annoying delays while disks spin up unexpectedly, and low battery life because of inappropriate device usage. Because the applications have direct knowledge of how the user is using the system to perform some function, this knowledge must penetrate into the power management decision-making system in order to prevent the kinds of user problems described above. This suggests that operating systems ought to provide application programming interfaces so that energy-aware applications may influence the scheduling of the systems resources.

The switching activity in a circuit is also a function of the present inputs and the previous state of the circuit. Thus it is expected that the energy consumed during execution of a particular instruction will vary depending on what the previous instruction was. Thus an appropriate reordering of instructions in a program can result in lower energy. Today, the cost function in most compilers is either speed or code size, so the most straightforward way to proceed is to modify the objective function used by existing code optimizers to obtain low-power versions of a given software program. The energy cost of each instruction must be considered during code optimization. An energy-aware compiler has to make a trade-off between size and speed in favor of energy reduction.

At the algorithm-level functional pipelining, re-timing, algebraic transformations, and loop transformations can be used [29]. The system's essential power dissipation can be estimated by a weighted sum of the number of operations in the algorithm that has to be performed. The weights used for the different operations should reflect the respective capacitance switched. The size and the complexity of an algorithm (e.g., operation counts, word length) determine the activity. Operand reduction includes common sub-expression elimination, dead code elimination, etc. Strength reduction can be applied to replace energy-consuming operations by a combination of simpler operations (for example, by replacing multiplications into shift and add operations).

8.4.4 Energy Reduction in Network Protocols

Up to this point we have mainly discussed the techniques that can be used to decrease the energy consumption of digital systems and focused on the computing components of a mobile host. In this subsection, we will discuss some techniques that can be used to reduce the energy consumption that is needed for the communication external of the mobile host.

We classify the sources of power consumption, with regard to network operations, into two types: (1) communication related and (2) computation related.

Communication involves usage of the transceiver at the source, intermediate (in the case of ad hoc networks), and destination nodes. The transmitter is used for sending control, route request, and response, as well as data packets originating at or routed through the transmitting node. The receiver is used to receive data and control packets some of which are destined for the receiving node and some of which are forwarded. Understanding the power characteristics of the mobile radio used in wireless devices is important for the efficient design of communication protocols.

The computation mainly involves usage of the CPU, main memory, the storage device, and other components. Also, data compression techniques, which reduce packet length, may result in increased power consumption due to increased computation. There exists a potential trade-off between computation and communication costs. Techniques that strive to achieve lower communication costs may result in higher computation needs, and vice-versa. Hence, protocols that are developed with energy efficiency goals should attempt to strike a balance between the two costs.

Energy reduction should be considered in the whole system of the mobile and through all layers of the protocol stack. The following discussion presents some general guidelines that may be adopted for an energy-efficient protocol design.

8.4.4.1 Protocol Stack Energy Reduction

Data communication protocols dictate the way in which electronic devices and systems exchange information by specifying a set of rules that should be a consistent, regular, and well-understood data transfer service. Mobile systems have strict constraints on the energy consumption, the communication bandwidth available, and are required to handle many classes of data transfer over a limited bandwidth wireless connection, including real-time traffic such as speed and video. For example, multimedia applications are characterized by their various media streams with different qualities of service requirements.

In order to save energy, an obvious mode of operation of the mobile host will be a sleep mode [34]. To support such mode the network protocols need to be modified. Store-and-forward schemes for wireless networks, such as the IEEE 802.11 proposed sleep mode, not only allow a network interface to enter a sleep mode but can also perform local retransmissions not involving the higher network protocol layers.

There are several techniques used to reduce the power consumption in all layers within the protocol stack. In Fig. 8.4, we list areas in which conservation mechanisms are efficient.

Collisions should be eliminated as much as possible within the media access layer (MAC) layer, a sub-layer of the data link layer, since they result in retransmissions. Retransmissions lead to unnecessary power consumption and to possibly unbounded delays. Retransmissions cannot be completely avoided in a wireless network due to the high error rates. Similarly, it may not be possible to fully eliminate collisions in a wireless mobile network. This is partly due to user mobility and a constantly varying set of mobiles in a cell.

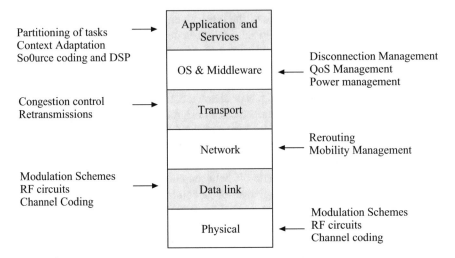

Fig. 8.4 Protocol stack of a generic wireless network, and corresponding areas of energy-efficient possible research

For example, new users registering with the base station may have to use some form of random access protocol. In this case, using a small packet size for registration and bandwidth request may reduce energy consumption. The EC-MAC protocol [34] is one example that avoids collisions during reservation and data packet transmission. This is the default mechanism used in the IEEE 802.11 wireless protocol in which the receiver is expected to keep track of channel status through constant monitoring. One solution is to broadcast a schedule that contains data transmission starting times for each mobile as in [34]. Another solution is to turn off the transceiver whenever the node determines that it will not be receiving data for a period of time.

Physical Layer

As shown in Fig. 8.4, the lowest level of the protocol stack is the physical layer. This layer consists of radio frequency (RF) circuits, modulation, and channel coding systems. At this level, we need to use an energy-efficient radio that can be in various operating modes (like variable RF power and different sleep modes) such that it allows a dynamic power management [35]. Energy can also be saved if it is able to adapt its modulation techniques and basic error correction schemes. The energy per bit transmitted or received tends to be lower at higher bit rates. For example, the WaveLAN radio operates at 2 Mb/s and consumes 1.8 W, or 0.9 J/bit.

A commercially available FM transceiver (Radiometrix BIM-433) operates at 40 kb/s and consumes 60 mW, or 1.5 J/bit. This makes the low bit-rate radio less efficient in energy consumption for the same amount of data. However, when a mobile has to listen for a longer period for a broadcast or wake-up from the base station, then the high bit-rate radio consumes about 30 times more energy than the low bit-rate radio. Therefore, the low bit-rate radio must be used for the basic signaling only, and as little as possible for data transfer. To minimize the energy consumption, but also to mitigate interference and increase network capacity, the transmit power on the link should be minimized, if possible.

Data Link Layer

The data link layer is thus responsible for wireless link error control, security (encryption/decryption), mapping network layer packets into frames, and packet retransmission. A sub-layer of the data link layer, the media access control (MAC) protocol layer, is responsible for allocating the time–frequency or code space among mobiles sharing wireless channels in a region.

In an energy-efficient MAC protocol the basic objective is to minimize all actions of the network interface, i.e., minimize on-time of the transmitter as well as the receiver. Another way to reduce energy consumption is by minimizing the number of transitions the wireless interface has to make. By scheduling data transfers in bulk, an inactive terminal is allowed to doze and power off the receiver as long as the network interface is reactivated at the scheduled time to transmit the data at full speed.

An example of an energy-efficient MAC protocol is E^2MaC [36]. The E2MaC protocol is designed to provide QoS to various service classes with a low energy consumption of the mobile. In this protocol, the main complexity is moved from the mobile to the base station with plenty of energy. The scheduler of the base station is responsible to provide the connections on the wireless link the required QoS and tries to minimize the amount of energy spent by the mobile. The main principles of the E^2MaC protocol are to avoid unsuccessful actions, minimize the number of transitions, and synchronize the mobile and the base station.

Network Layer

The network layer is responsible for routing packets, establishing the network service type, and transferring packets between the transport and link layers. In a mobile environment this layer has the added responsibility of rerouting packets and mobility management. Errors on the wireless link can be propagated in the protocol stack. In the presence of a high packet error rate and periods of intermittent connectivity of wireless links, some network protocols (such as TCP) may overreact to packet losses, mistaking them for congestion. TCP responds to all losses by invoking congestion control and avoidance algorithms. These measures result in an unnecessary reduction in the link's bandwidth utilization and increases in energy consumption because it leads to a longer transfer time.

The limitations of TCP can be overcome by a more adequate congestion control during packet errors. These schemes choose from a variety of mechanisms to improve end-to-end throughput, such as local retransmissions, split connections, and forward error correction.

A comparative analysis of several techniques to improve the end-to-end performance of TCP over lossy, wireless hops is given [37]. These schemes are classified into three categories: end-to-end protocols, where loss recovery is performed by the sender; link-layer protocols that provide local reliability; and split-connection protocols that break the end-to-end connection into two parts at the base station. The results show that a reliable link-layer protocol with some knowledge of TCP provides good performance, more than using a split-connection approach. Selective acknowledgment schemes are useful, especially when the losses occur in bursts.

OS and Middle-Ware Layer

The operating system and middle-ware layer handles disconnection, adaptively support, and power and QoS management within wireless devices. This is in addition to the conventional tasks such as process scheduling and file system management. To avoid the high cost, in terms of performance, energy consumption or money of wireless network communication is to avoid use of the network when it is expensive by predicting future access and fetching necessary data when the network is cheap. In the higher level, protocols of a communication system caching and scheduling can be used to control the transmission of messages. This works in particular well when the computer system has the ability to use various networking infrastructures (depending on the availability of the infrastructure at a certain locality), with varying and multiple network connectivities and with different characteristics and costs. True prescience, of course, requires knowledge of the future. Two possible techniques, LRU caching and hoarding, are for example present in the Coda cache manager. A summary of other software strategies for energy efficiency is presented in [28, 38].

8.5 Chapter Summary

Power dissipation continues to be a primary design constraint in single- and multicore-based systems. Increasing power consumption not only results in increasing energy costs, but also results in high die temperatures that affect chip reliability, performance, and packaging cost.

This chapter has investigated a number of energy-aware design techniques that can be used into complex multicore systems. In particular, this chapter covered techniques used to design energy-aware systems at the technology, logic, and system levels. The vast majority of the techniques used in the system architectures are derived from existing uni-processor energy-aware systems.

References

1. R. Kravets, P. Krishnan, Application driven power management for mobile communication Springer Science. Wirel. Netw. **6**(4), 263–277 (2000)
2. J. Lorch, Modeling the effect of different processor cycling techniques on power consumption, Performance evaluation group technical note 179, in *ATG Integrated Sys, Apple Computer*, 1995

3. T. Martin, Balancing batteries, power and performance: system issues in CPU speed-setting for mobile computing, Ph.D. Dissertation, Carnegie Mellon University, Department of Electrical and Computer Engineering, Aug, 1999
4. G.F. Welch, A survey of power management techniques in mobile computing operating systems. ACM SIGOPS Oper. Syst. Rev. **29**(4), 47–56 (1995)
5. J.M. Rulnick, N. Bambos, Mobile power management for maximum battery life in wireless communication networks, in *Proceedings of IEEE INFOCOM 96*, 1996
6. F.N. Najm, A survey of power estimation techniques in VLSI circuits. IEEE Trans. VLSI Syst. **2**(4), 44–55 (1994)
7. M. Pedram, Power minimization in IC design: principles and applications. ACM Trans. Des. Autom. Electron. Syst. **1**(1), 3–6 (1996)
8. S. Borkar, Design challenges of technology scaling. IEEE Micro. **19**(4), 23–29 (1999)
9. Semiconductor industry association: the national technology roadmap for semiconductors: technology needs, Sematche Inc. (Austin, USA, 1997), http://www.sematech.org
10. Y. Ye. S. Borkar, V. De. A new technique for standby leakage reduction in high-performance circuits, in *Symposium on VLSI Circuits,* Honolulu, Hawaii, 40–41 1998
11. L. Benini, G. de Micheli, System-level power optimization: techniques and tools, in *Proceedings Intéz Symposium Low-Power Electronics Design*, San Diego, CA, 288–293 1999
12. L. Benini, G. de Micheli, E. Macii, Designing low-power circuits: practical recipes. IEEE Circuits Syst. Mag. **1**, 6–25 (2001)
13. A. Ben Abdallah, S. Kawata, T. Yoshinaga, M. Sowa, Modular design structure and high-level prototyping for novel embedded processor core, in *Proceedings of the 2005 IFIP International Conference on Embedded And Ubiquitous Computing (EUC'2005),* Nagasaki, 340–349 Dec 6–9 2005
14. V. Tiwari, S. Malik, Guarded evaluation: pushing power management to logic synthesis/design. IEEE Trans. Comput. Aided Des. Integr. Circ. Syst. **17**(10), 1051–1060 (1998)
15. A. Abnous, J. Rabaey, Ultra-low-power domain-specific multimedia processors, in *(Proceedings of the IEEE VLSI Signal Processing Workshop, IEEE press*, 459–464 1996
16. J. Liang et al., An architecture and compiler for scalable on-chip communication. IEEE Trans. VLSI Syst. **12**(7), 711–726 (2004)
17. J. Rabaey, L. Guerra, R.Mehra, Design guidance in the power Dimension, in *Proceedings of the ICASSP*, 1995
18. C.L. Su, M. Alvin Despain, cache designs for energy efficiency, in *Proceedings of the 28th Hawaii International Conference on System Science*, 1995
19. H. Mehta, R. M. Owens, M.J. Irwin, R. Chen, D. Ghosh, Techniques for low energy software, in *Internatzonal Symposzum of Low Power Electronics and Deszgn, IEEE/ACM*, 72–75 1997
20. F. Doughs, P. Krishnan, B. Marsh, Thwarting the power hungry disk, in *Proceedings of the 1991 Winter USENIX Conference*, 1994
21. F. Douglis, F. Kaashoek, B. March, B. Caceres, K. Li, J. Tauber, Storage alternative for mobile computers, in *Proceedings of the first USENIX Symposimum on Operating Systems Design and Implemnetation*, 1994
22. K. Li, R. Kumpf, P. Horton, T. Anderson, A quantitative analysis of disk drive power management in portable computers, in *Proceedings of the 1994 Winter USENIX*, 1994
23. P. Erik, P. Harris, W. Steven, E.W. Pence, S. Kirkpatrick, Technology directions for portable computers. in *Proceedings of the IEEE*, 83b(4) 63–57 1995
24. K. Govil, E. Chan, H. Wasserman. Comparing algorithms for dynamic speed-setting of a low-power cpu, In *First ACM International Conference on Mobile Computing and Networking (MOBICOM)*, 1995
25. J.R. Lorch, A.J. Smith. Reducing processor power consumption by improving processor time management in a single user operating system, in *Second ACM International Conference on Mobile Computing and Networking (MOBICOM)*, 1996
26. M. Weiser, B. Welch, A. Demers, S. Shenker. Schedlibng for reduced cpu energy. in*Proceedings of the First Symposium on Operating System Design and Implementation (OSDI)*, 1994

27. J. Lorch, A complete picture of the energy consumption of a portable computer. Master's thesis, Department of Computer Science, University of California at Berkeley, 1995
28. J.R. Lorch, A.J. Smith, Software strategies for portable computer energy management. IEEE Pers. Commun. 5(3), 60–73 (1998)
29. V. Tiwari, S. Malik, A. Wolfe, Power analysis of embedded software: a first step towards software power minimization. IEEE Trans. Very Large Scale Integr. 2(4), 437–445 (1994)
30. F. Wolf, *Behavioral Intervals in Embedded Software: Timing and Power Analysis of Embedded Real-Time Software Process* (Kluwer Academic Publishers, The Netherlands, 2002). ISBN: 1-4020-7135-3
31. Intel Corporation. Mobile power guidelines (2000), ftp://download.intel.com/design/mobile/intelpower/mpg99r1.pdf, Accessed Dec 1998
32. Intel Corporation: Mobile intel pentium III processor in BGA2 and micro-PGA2 packages, revision 7.0, (2001)
33. P.J.M. Havinga, G.J.M. Smit, Energy-efficient wireless networking for multimedia applications, in wireless communications and mobile computing. Wirel. Commun. Mob. Comput. 1, 165–184 (2001)
34. K.M. Sivalingam, J.C. Chen, P. Agrawal, M. Srivastava, Design and analysis of low-power access protocols for wireless and mobile ATM networks. Wirel. Netw. 6(1), 73–87 (2000)
35. F. Akyildiz, S. Weilian, S. Yogesh, E. Cayirci, A Survey on sensor networks. IEEE Commun. Mag. 40(8), 102–114 (2002)
36. P.J. Havinga, G. Smit, M. Bos, Energy-efficient wireless ATM design, in (Proceedings wmATM), 2–4 June 1999
37. H. Balakrishnan, V. Enkata, N. Padmanabhan, A comparison of mechanisms for improving tcp performance over wireless links. IEEE/ACM Trans. Netw. 5(6), 756–769 (1997)
38. J. Kistler, Disconnected operation in a distributed file system, Ph.D. thesis, Carnegie Mellon University, School of Computer Science, 1993

Chapter 9
Real Deign of Embedded Multicore SoC for Health Monitoring

Abstract Recent technological advances in wireless networking, embedded micro-electronics and the Internet allow computer and biomedical scientists to fundamentally modernize and change the way health care services are deployed. Thus, changes and new services are urgently needed to help cope with the imminent crisis in the health care systems caused by current demographic, social, and economic trends in many countries. Electrocardiography (ECG) is an interpretation of the electrical activity of the heart over time captured and externally recorded by electrodes. An effective approach to speed up this and other biomedical operations is to integrate a very high number of processing elements in a single chip so that the massive scale of fine-grain parallelism inherent in several biomedical applications can be exploited efficiently. In this chapter, we present a case study of a real hardware and software design of a multicore SoC architecture targeted for elderly health monitoring.

9.1 Introduction

The world population over age 65 is expected to more than double from 357 million in 1990 to 761 million in 2025 [1]. These statistics clearly underscore the need for more scalable and affordable health care solution. Despite the decreased mortality rate, heart disease and associated complications is one of the main causes of death around the world. Detection of irregularities in the rhythms of the heart is a growing concern in medical research.

Embedded health monitoring systems are approaches to deal with such problems. However, development of such systems faces a number of challenging tasks since they need to often address conflicting requirements for performance, size, and accuracy. In health monitoring applications, a wide range of parameters must be available and processed. Thus, multiple tasks must be performed in order to obtain accurate diagnosis. In most cases, complex computations are required because of the applied detection algorithm. When real time diagnosis are required, a single medium-performance processor core might have problems dealing with all tasks.

© Springer Nature Singapore Pte Ltd. 2017
A. Ben Abdallah, *Advanced Multicore Systems-On-Chip*,
DOI 10.1007/978-981-10-6092-2_9

Electrocardiography is an essential practice in heart medicine. It faces computational challenges, especially when 12 or more lead signals are to be analyzed in parallel, in real-time, and under increasing sampling frequencies. Another challenge is the analysis of huge amounts of data that may grow to days of recording.

As we described in Chap. 3, MCSoCs are high performance devices that incorporate multiple building blocks from multiple sources. An MCSoC may contain general or special purpose fully programmed processors, co-processors, DSPs, dedicated hardware, memory blocks, etc. These systems are becoming a common design alternative in portable devices because it is possible to manufacture a silicon chip including only necessary elements. Certain medical applications require devices capable of providing very accurate information about monitored patients in places where complex clinical systems are not available and parameters such as electrocardiogram characteristics are important to be determined.

In this chapter, we present a case study of a real hardware and software design of a multicore SoC architecture, named BANSMOM, targeted for biomedical applications. The ultimate goal of this multicore system is to monitor elderly people and promote well-being by introducing smart in-body sensors that allow medical professionals to initiate interventions in the home environment. Although the designed system can be slightly modified and used for various biomedical applications, we focus here only on ECG processing and real-time monitoring. The complete system consists of hardware and software components. The hardware part consists of the multicore SoC platform. While the software part consists of a so called *Period-Peak-Detection (PPD)* algorithm and a real-time interaction interface.

9.1.1 Electrocardiography and Heart Diseases

Electrocardiography, which is also called ECG/EKG, is generally used to record the electrical impulses which immediately precede the contractions of the heart muscle. This method causes no discomfort to a patient and is often used for diagnosing heart disorders such as coronary heart disease, pericardia or inflammation of the membrane around the heart, heart muscle disease, arrhythmia and coronary thrombosis, etc. When using this technique, doctors connect electrodes to the chest, wrist and ankles that are connected to a recording machine. This machine displays, then, the electrical activity in the heart as a trace on a rolling graph or screen. Using the electrocardiography any abnormalities are revealed to the doctor.

This technique can be taken at a doctor's office, hospital or even at home and will provide your doctor with a 24 h record of the patients heart activity from a tape recorder that is worn by the patient. The Doctor or a medical staff can look at the printed graph to see if the heart chambers are contracting with complete regularity which indicates a normal rhythm. If the contractions of the lower heart chambers are extremely irregular this could indicate ventricular fibrillation. When the upper and lower heart chambers are beating independently this could indicate a complete heart blockage. If the upper heart chambers are beating fast and irregular, this can indicate arterial fibrillation.

The electrocardiography is a painless and quick procedure. The electrical impulses in the heart are recorded and amplified on a moving strip of paper. Small metal electrodes are placed on the skin of the patient to measure the flow and direction of the electrical currents in the heart during each heart beat. Each of the electrodes are connected by a wire to a machine that will produce what is called a tracing for each electrode. This tracing represents a particular view or what is called lead of the heart's electrical patterns. In most cases any person who is suspected of having heart disease will have an ECG taken by their doctor. This will aid the doctor in identifying a number of heart problems.

An electrocardiography produces waves that are known as: P, Q, R, S, T and U waves which gives each part of the ECG an alphabetical designation. Figure 9.1 shows an example of a typical ECG wave. As the heart beat begins with an impulse from the senatorial node, the impulse will first activate the upper chambers of the heart or atria and produce the P wave. Then the electrical current will flow down to the lower chambers of the heart or ventricles producing the Q, R and S waves. As the electrical current spreads back over the ventricles in the opposite direction it will produce the T waves. Using this technique doctors can determine where in the heart abnormal rhythms start which allows them to begin to determine the cause.

Many ECG analysis methods use the three peaks Q, R, S and the corresponding intervals between these three peaks. In biomedical terms, this interval from Q to R to S is known as the QRS complex [2, 3]. The well known QRS Pan-Tompkins algorithm locates R-peaks in the ECG signal and calculates the heart period [4].

A number of other research efforts focus on hardware implementations of health monitoring systems. For example, Christos presented a hardware implementation of the Pan and Tompkins QRS complex [5].

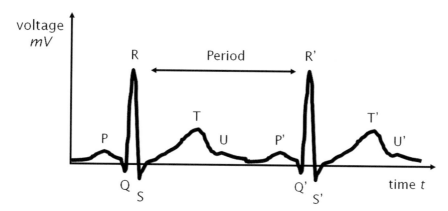

Fig. 9.1 A typical ECG wave

9.2   Application Specific Digital Signal Processing

Before we start describing the system architecture, let us first review some basic knowledge about digital signal processing (DSP). DSP has been with us for decades now with some astronomical development in the area over the years. The world of science and engineering is filled with signals: images from remote space probes, voltages generated by the heart and brain, radar and sonar echoes, seismic vibrations, and countless other applications. Digital Signal Processing is the science of using computers to understand these types of data. This includes a wide variety of goals: filtering, speech recognition, image enhancement, data compression, neural networks, and much more. DSP is one of the most powerful technologies that will shape science and engineering in the twenty-first century.

DSP is the processing of signals by digital means. A signal in this context can mean a number of different things. Historically the origins of signal processing are in electrical engineering, and a signal here means an electrical signal carried by a wire or telephone line, or perhaps by a radio wave. More generally, however, a signal is a stream of information representing anything from stock prices to data from a remote-sensing satellite. A digital signal consists of a stream of numbers, usually (but not necessarily) in binary form. The processing of a digital signal is done by performing numerical calculations.

One of the important factors that enhanced DSP performance is the evolution of the processor architecture. As the computing needs of each processor type grew rapidly, the first response of the computer architects and semiconductor companies was to increase the speed of the single processor. Higher performance was mainly achieved by refining manufacturing processes to improve the operating speed. However, this method requires finding solutions for increased leakage power and other problems, making it unable to keep pace with the current rate of evolution.

After several decades of single processor core devices production, major CPU makers, such as Intel and AMD, decided to switch to multicore processor chips because it was found that several smaller processor cores running at a lower frequency can perform the same amount of work without consuming as much energy and power. More precisely, this shift started when Intel's hardware engineers launched the Pentium 4; at that time, they expected single processor chip to scale up to 10 GHz or even more using advanced process technologies below 90 nm. However, they did not really achieve their expectation since the fastest processor never exceeded 4 GHz. As a result, the trends followed by all major hardware makers is to use a higher number of slower logic gates, building parallel devices made with denser chips that work at low clock speed. One of the areas that takes advantage from the evolution of the processor architecture and digital signals processing is the embedded systems targeted for health monitoring. Such systems require small devices capable to process a bio-medical data at real time and with the minimum energy use.

Many micro-watt power processors have been proposed to improve the processing efficiency for the possible application to Bio Signal Processing. The first group is the general purposed processor. They have developed for low power operation.

Yet, they still require the long operating time, which is the important factor of the energy consumption. Thus, the application specific processor rather than general purpose processor has been developed. Even though it consumes more power than the general purposed processors, the operating time can be reduced remarkably due to the dedicated hardware and instructions. Thus, if the application is clearly defined such as the Bio Signal Processing, it becomes very attractive to improve the energy efficiency.

Recently, with the increase of the interests in the healthcare, the need for the ambulatory arrhythmia monitoring system has been rising exponentially. The monitoring system records ECG signal continuously in ambulatory condition for a sizable time like several hours. The system transmits the record data to the user or the healthcare center like hospital when the alert ECG signal is detected or the recording period is finished. In order to monitor and analyze the ECG signal, the functions operated at the clinical instrument such as signal sensing and the classification should be integrated into the light-weight, ambulatory monitoring system. The most important requirements for the ambulatory monitoring system are ultra low energy operation for the long battery life time and a small footprint for wearability. In general, since the highest energy consuming parts are the memory transaction blocks and the wireless communication blocks than the processing block, the data processing as much as possible before transmission is the most efficient method to reduce the total system energy consumption. However, the development of such systems faces several challenging tasks since they need to often address conflicting requirements for performance, size, and accuracy. In addition, Electrocardiography (ECG) may face many computational challenges, especially when 12 or more lead signals should be analyzed in parallel, at real-time, and under high sampling frequencies. A number of recent research efforts focus on hardware implementations and the real time visualization of the ECG signals *Christos* presented a hardware implementation of the Pan and Tompkins QRS detection algorithm [5]. The system achieved a speed up of 250% compared to the software implementation. *Yutana Jewajinda* presented an FPGA-based Online-learning for ECG Signal Classification [6], where the complete ECG signal classification can be implemented in hardware. Another challenge is the analysis of huge amounts of data that may grow depending on the recording range time and also the Visualization outputs of such systems on an environment that can be consulted at real time. Recent techniques deployed for monitoring heart activity is the 12-leads ECG which uses data coming from twelve ECG leads serially. The leads produce huge amounts of data, especially when used for a long number of hours. The monitoring part is also crucial for the real time diagnosis. This importance came from the fact that in order to make clinical studies or heart diseases diagnosis, researchers and doctors are required to be in immediate proximity to patients which is not always practical.

Traditionally, personal medical monitoring systems, such as ambulatory electrocardiography devices [7], have been used to record data. Data processing and analysis are performed off-line, making such devices impractical for continual monitoring and early detection of medical disorders, especially for patients needing immediate medical interventions.

A lot of research have been conducted to perform the transfer and the diagnosis of the ECG signals at real time, Farah Magrabi et al. [8] presented a Web Based Longitudinal ECG Monitoring, Jun Dong et al. [9] presented A Remote Diagnosis Service Platform for Wearable ECG Monitors where they present a patient location independent and continuous ECG monitoring and diagnosis system.

9.2.1 Analog and Digital Signals

In many cases, the signal of interest is initially in the form of an analog electrical voltage or current, produced for example by a microphone or some other type of transducer. In some situations, such as the output from the readout system of a compact disc player, the data is already in digital form. An analog signal must be converted into digital form before DSP techniques can be applied. An analog electrical voltage signal, for example, can be digitized using an electronic circuit called an analog-to-digital converter (ADC). This generates a digital output as a stream of binary numbers whose values represent the electrical voltage input to the device at each sampling instant.

9.2.2 Signal Processing

Signals commonly need to be processed in a variety of ways. For example, the output signal from a transducer may well be contaminated with unwanted electrical noise. The electrodes attached to a patient's chest when an ECG is taken measure tiny electrical voltage changes due to the activity of the heart and other muscles. The signal is often strongly affected by mains pickup due to electrical interference from the mains supply. Processing the signal using a filter circuit can remove or at least reduce the unwanted part of the signal. Increasingly nowadays, the filtering of signals to improve signal quality or to extract important information is done by DSP techniques.

9.2.3 Analog to Digital Conversion

Signals in the real world are analog: light, sound, heart signal. So, real-world signals must be converted into digital, using a circuit called analog to digital conversion, before they can be manipulated by digital equipment. Scanning a picture for example with a scanner is doing is an analog-to-digital conversion: The scanner is taking the analog information provided by the picture (light) and converting into digital. When voice is recorded or a VoIP solution is used on the computer, an analog-to-digital conversion takes place to convert voice, which is analog, into digital information.

Digital information is not only restricted to computers. For talk on the phone, for example, voice is converted into digital since voice is analog and the communication between the phone switches is done digitally. When an audio CD is recorded at a studio, analog-to-digital conversion is taking place, converting sounds into digital numbers that will be stored on the disc.

To get the analog signal back, the opposite conversion digital-to-analog, which is done by a circuit called DAC (Digital-to-Analog Converter) is needed. Playing an audio CD, what the CD player is doing is reading digital information stored on the disc and converting it back to analog so that the music can be heard. There are some basic reasons to use digital signals instead of analog, noise being the number one. Since analog signals can assume any value, noise is interpreted as being part of the original signal. Digital systems, on the other hand, can only understand two numbers, zero and one. Anything different from this is discarded.

Basically Analog-to-digital conversion is an electronic process in which a continuously variable (analog) signal is changed, without altering its essential content, into a multi-level (digital) signal. The input to an analog-to-digital converter (ADC) consists of a voltage that varies among a theoretically infinite number of values. Examples are sine waves, the waveforms representing human speech, heart signals and the signals from a conventional television camera.

The output of the ADC, in contrast, has defined levels or states. The number of states is almost always a power of two – that is, 2, 4, 8, 16, etc. The simplest digital signals have only two states, and are called binary. All whole numbers can be represented in binary form as strings of ones and zeros. Digital signals propagate more efficiently than analog signals, largely because digital impulses, which are well-defined and orderly, are easier for electronic circuits to distinguish from noise, which is chaotic. This is the chief advantage of digital modes in communications. Computers "talk" and "think" in terms of binary digital data; while a microprocessor can analyze analog data, it must be converted into digital form for the computer to make sense of it.

9.3 Period-Peak ECG Detection Algorithm

After we reviewed some basics about DSPs, we will now describe in this section a so called Period-Peak Detection Algorithm (PPD) for processing real signal - ECG signals. The PPD algorithm first detects the period and then looks for all peaks [10, 11]. This is the fundamental idea which differ from the known approaches that first find peaks and then period.

The PPD algorithm detects period before finding peaks because there is a high degree of randomness in the ECG signals. The randomness makes finding peaks an erroneous process. Figure 9.2 illustrates an example of faulty ECG analysis.

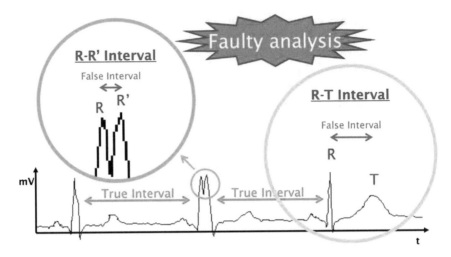

Fig. 9.2 Faulty ECG Analysis

What we would get by doing this is the level of correlation these signals have. PPD algorithm computes several parameters: heart period, typical peaks (P, Q, R, S, T, and U), and inter-peak time spans (R-R interval). Peak height and inter-peak time ranging outside normal values, indicating different kinds of diseases, are also detected with the PPD algorithm.

PPD consists of two processing flows (see Fig. 9.3): one that detects period using the autocorrelation function, and another one that detects the number, amplitude and time interval of all peaks.

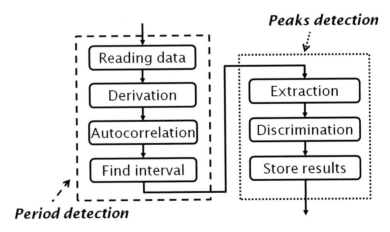

Fig. 9.3 PPD algorithm processing flow

9.3.1 Period Detection

As indicated in Fig. 9.3, the PPD algorithm's *period detection* phase consists of 4 steps: (1) Data reading, (2) Derivation, (3) Autocorrelation, and (4) Finding intervals. The *derivation phase* finds the discrete derivative of the ECG signal. The derivative function is the best function we can run and which can aid in amplifying signal peaks; therefore, after reading the data of a signal y with samples from memory, a very helpful step is to calculate the derivative of the signal $y(t)$. Equation 9.1 is the derivative function used by the PPD algorithm in the *period detection* phase.

$$\frac{\partial y}{\partial t}(t) \approx \frac{y[n+1] - y[n]}{(n+1) - n} = y[n+1] - y[n] \tag{9.1}$$

With this derivative, a given peak will be amplified relative to the samples before it, and if the value of $y[n]$ and $y[n+1]$ are near each other (i.e. no peaks) then the difference will look relatively small on the new derivative graph. The advantage of taking the derivative is that the fluctuations taking place in the signal, especially those around the peaks, would be reduced to a near-zero-value. In addition, performance overhead associated with derivative calculation of the ECG signal is negligible compared to the rest of the algorithm.

The *autocorrelation* step finds the period of ECG signals. This step uses autocorrelation function (ACF) defined by Formula 9.2. This ACF is a statistical method used to measure the degree of association between values in a single series separated by some lags. The fixed length ACF is defined by Formula 9.3. By running the ACF on the function y over the recorded data sample, we can easily get the coefficients of the ACF.

$$R_y[k] = \sum_{n=-\infty}^{n=\infty} y[n] \times y[n-k] \tag{9.2}$$

$$R_y[L] = \sum_{n=0}^{N} y[n] \times y[n-L] \tag{9.3}$$

where, R_y is the autocorrelation function, $y[n]$ is the filtered ECG signal. L is a positive natural number related to the number of times needed for the calculations to get the period; it is the same as the number of lags of the autocorrelation. Finally, the *finding interval* step finds interval point in ECG signals based on the results of the autocorrelation step.

The *period detection* detailed computation steps are shown in Fig. 9.4. As it is shown in the above flow-chart, the actual *period detection* phase consists of seven steps. Figures 9.6, 9.7, 9.8 and 9.9 describe the detailed computations of the main steps (Fig. 9.5).

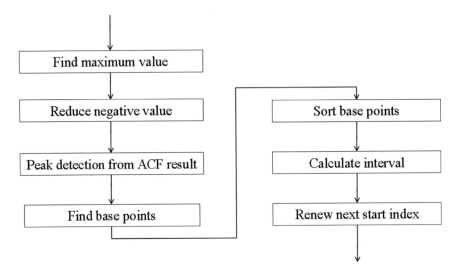

Fig. 9.4 Period detection computation details

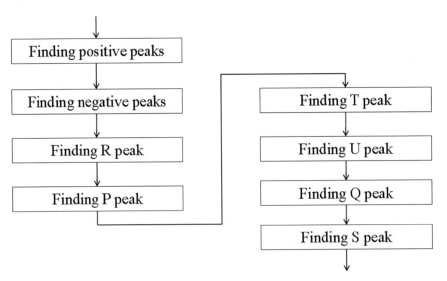

Fig. 9.5 Peaks detection computation details

9.3.2 Peaks Detection

As shown in Fig. 9.3, the second phase of the PPD algorithm is the *peaks detection* phase and consists of 3 main steps: Extraction, Discrimination, and Store results. The *extraction* step discriminates significant peaks from calculated interval information in *period detection* phase. The *discrimination* step finds 6 peak-points (P, Q, R, S,

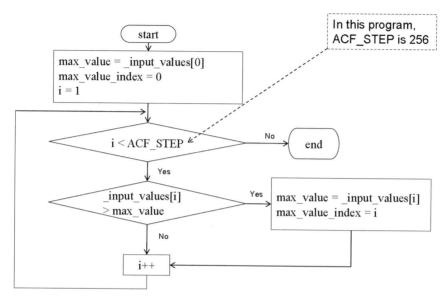

Fig. 9.6 Period detection: finding maximum value algorithm. The autocorrelation step ACF_STEP is set 256

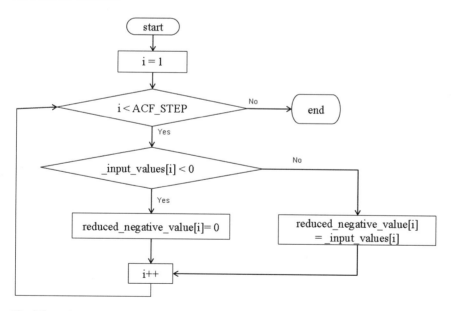

Fig. 9.7 Period detection: reduce negative value algorithm

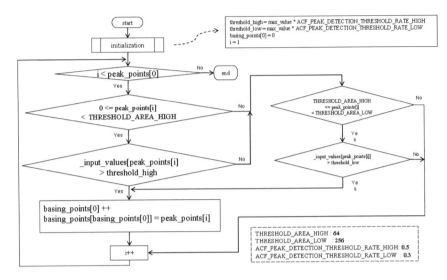

Fig. 9.8 Period detection: find base points

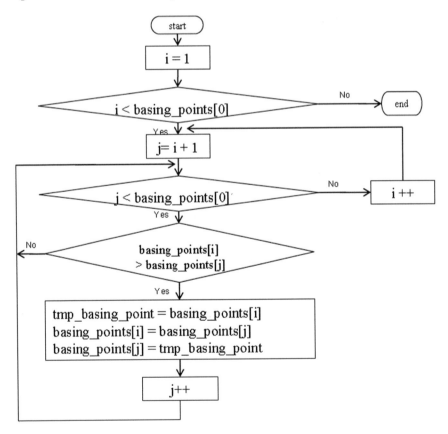

Fig. 9.9 Period detection: sort base points

T and U) from the extracted peaks [1, 12]. Finally, the *store* step stores interval and peak information in the buffer. The *peaks detection* detailed computation is shown in Fig. 9.5.

9.4 Multicore SoC Hardware Design

Figure 9.10 shows the block diagram of the multicare SoC system architecture that was designed in hardware. As shown in this figure, the processing of a given ECG signal from one lead is performed in four major phases: (1) signal reading, (2) filtering, (3) analysis, and (4) display. In the remaining of this section, we will explain these processing stages in detail.

9.4.1 Signal Reading

First of all we have to note that the number of data inputs from the external sensors (leads) is extensible to 15 leads or more (see Fig. 9.10). The size of data is included in data read from the sensors. Analog Digital Converter (ADC) converts analog data into digital data so a contiguous ECG signal is converted into a discrete ECG signal. As a result, filter processing and analysis processing can be easily done.

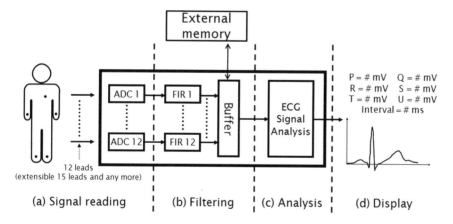

Fig. 9.10 High-level view of the BANSMOM system architecture

9.4.2 *Filtering*

The band-pass filter reduces the influence of muscle noise, 60 Hz interference, baseline wander, and T-wave interference. This filter cascaded the low-pass and high-pass filters described below to achieve a 3 dB passband from about 5–11 Hz.

A low pass filter only allows low frequency signals from 0 Hz to its cut-off frequency, fc point to pass while blocking those any higher. The difference equation of the filter is shown in Formula 9.4. Where T is the sampling period, the cutoff frequency is about 11 Hz.

$$y(nT) = 2y(nT - T) - y(nT - 2T) + x(nT)$$
$$-2x(nT - 6T) + x(nT - 12T) \qquad (9.4)$$

The other type of filtering is called high-pass filter. Its design is based on subtracting the output of a first-order low-pass filter from an all-pass filter (i.e., the samples in the original signal). The difference equation of the filter is shown in the Formula 9.5. Where the cutoff frequency is about 5 Hz.

$$y(nT) = 32x(nT - 16T) - [y(nT - T)$$
$$+x(nT) - x(nT - 32T)] \qquad (9.5)$$

9.4.2.1 Noise Filtering

Noise filtering uses a bandpass filter that is based on the Finite Impulse Response (FIR) filter. The bandpass filter reduces the influence of muscle noise, 50 Hz interference, baseline wander, and T-wave interference. Digital filters process digitized or sampled signals. A digital filter computes a quantized time-domain representation of the convolution of the sampled input time function and a representation of the weighting function of the filter. They are realized by an extended sequence of multiplications and additions carried out at a uniformly spaced sample interval. The digitized input signal is mathematically influenced by the DSP program. These signals are passed through structures that shift the clocked data into summers (adders), delay blocks and multipliers. These structures change the mathematical values in a predetermined way: the resulting data represents the filtered or transformed signal.

Digital filters are a very important part of DSP. Filters have two uses: signal separation and signal restoration. Signal separation is needed when a signal has been contaminated with interference, noise, or other signals. For example, imagine a device for measuring the electrical activity of a baby's heart (EKG) while still in the womb. The raw signal will likely be corrupted by the breathing and heartbeat of the mother. A filter might be used to separate these signals so that they can be individually analyzed. Signal restoration is used when a signal has been distorted in some way. For example, an audio recording made with poor equipment may be filtered to better represent the sound as it actually occurred.

Finite Impulse Response (FIR) filter is a basic type of digital filter. FIR filters have no non-zero feedback coefficient in the general form of the digital filter difference equation. That is, the filter has only zeros, and once it has been excited with an impulse, the output is present for only a finite N number of computational cycles. The FIR filter uses noise rejection and waveform extraction for ECG algorithm. The data from analog/digital converter is finite and is a discrete digital signal; therefore, BANSMOM system uses FIR filter. This filter is popular among liner digital filters and the most safety in another filter within finite data. The FIR filter is composed of three parts: delay element, multiplier, and adder. Formula 9.7 is the difference equation for FIR filter that is defined by the relationship between input signal and output signal.

$$y[n] = a_0 x_n + a_1 x_{n-1} + \cdots + a_N x_{n-N} \tag{9.6}$$

$$= \sum_{i=0}^{N} a_i x_{n-i} \tag{9.7}$$

N is filter order that corresponds to the number of taps. xn are current or previous filter inputs. $y[n]$ is the current filter output. ai are the filters coefficients that correspond to impulse response. A FIR filter works by multiplying an array of the most recent n data samples by an array of constants, and summing the elements of the resulting array. The filter then inputs another sample of data and repeats the process.

9.4.3 Data Processing

Correlation is calculated between the acquired segment and a pattern which has been previously obtained. For each analyzed patient, the pattern segment is assumed to contain regular ECG signals where a signal's QRS complex is contained so that any further ECG pulses can be correlated with it. High correlation values correspond to pulse detection. Pattern extraction is performed by the main processor. Once obtained, it is transferred and stored to the off-chip memory. The off-chip memory starts to receive data samples directly from the ADC. Data samples are stored in a Queue (FIFO). Correlation pattern is calculated, and pulse alignment is evaluated. When the input signals in the Queue are aligned with the pattern, a high correlation value will be obtained, and a signal indicating the presence of a new pulse is generated. External monitor outputs the analyzed results. The output data is peaks of each typical wave (P, Q, R, S and T), heart rate and entire waveform (see Figs. 9.13 and 9.16).

9.4.4 Processor Core

For Master and Slave cores, we adopted the Nios II soft-core which is a 32-bit embedded-processor architecture designed by Altera [13]. Nios II incorporates many enhancements over the original Nios architecture, making it more suitable for a wider range of embedded computing applications, from DSP to system-control. The Nios II architecture is a RISC style architecture. The soft-core nature of the Nios II processor lets the system designer specify and generate a custom Nios II core, tailored for his or her specific application requirements. Of course, system designers can extend the Nios II's basic functionality by adding a predefined memory management unit, or defining custom instructions and custom peripherals (Fig. 9.11).

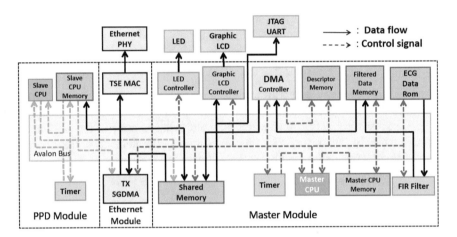

Fig. 9.11 Prototyped multicore SoC block diagram

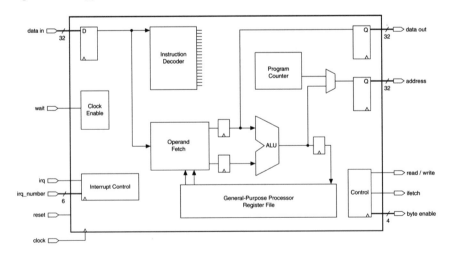

Fig. 9.12 Nios II core architecture block diagram

Nios II processor is designed for 5-stage pipelines, with separate data and instruction Harvard structure. Nios II has its own dedicated architecture and instruction set to support 32 bits hardware multiplication and division instructions. It has 32 general purpose registers. Users can also customize up to 256 instructions according to their needs. Figure 9.12 shows a simple block diagram of the Nios II core. Reader can reefer to on-line literature [14] for more details.

9.5 Real-Time Monitoring Interface Design

The monitoring part is crucial for the real time diagnosis with BANSNOM system. The existing methods of ECG monitoring are characterized by a manually-intensive work flow for data acquisition, formatting and visualization. Besides, they are most often relying on multiple serial processes and several software packages.

The developed Real-Time Interaction (RTI) interface is a robust web based application. This feature allows the user to deal with the high requirements of bio-medical data monitoring, such as the real time constraint in addition to the interaction and the synchronization issues between the medical staff side and the patient side.

The RTI tool uses PHP [15] as a server side and Mysql [16] as a data base management system. The ECG waveform is displayed with a library based on Javascript. Moreover, the dynamic update is ensured by Ajax technology to allow the user to interact directly with the data coming from the local storage without needing to refresh the page. The incoming data is stored in a dedicated table in Mysql database. Each node has its own table that contains all data classified by recording date. This classification allows the medical staff to visualize any specific data at any desired date and time. In addition, by building the RTI tool as a web application, one can improve the mobility of the monitoring task without distributing it or installing any specific software. Thus, the medical staff can consult the ECG data at real-time from anywhere through an Internet browser. In other words, the user/doctor can easily interact with patients at any time as long as the Internet connection is established (Figs. 9.13 and 9.14).

9.5.1 Data Capturing

Theoutput of the processing part is the coordinates of each peak for each corresponding node. To continue on the same way of mobility, the processed data coming from BANSMOM node(s) are transmitted to the database through the Internet and the corresponding table in MySQL is updated automatically. Figures 9.15 and 9.16 show BANSMOM system running snapshot and the external monitoring interface respectively.

```
*******************************************************************
*              The University of Aizu                    *
*         Embedded Multicore System Research             *
*           -- AIZUDAI BANSMOM Project --                *
*                      Author:Yasuyoshi Haga *
*              Last Update Date:Jan 22, 2010.    *
******* Period-Peaks Detection Processing Report ********
Sample Data:MIT-BIH Normal Sinus Rhythm Database No.16483 from PhysioBank
Filter lag:0.195s

*** Start of Processing

Range of processing: 0.000s - 2.000s
  Interval:0.625s [0.000s - 0.625s]
  PPP=0.047s, PPV=0.008mV
  QPP=0.086s, QPV=-0.292mV
  RPP=0.125s, RPV=0.492mV
  SPP=0.164s, SPV=-0.309mV
  TPP=0.344s, TPV=0.176mV
  UPP=0.602s, UPV=0.080mV
  R-R Interval:0.125s [0.000s - 0.125s]

  Interval:0.617s [0.625s - 1.242s]
  PPP=0.672s, PPV=0.003mV
  QPP=0.711s, QPV=-0.280mV
  RPP=0.750s, RPV=0.455mV
  SPP=0.797s, SPV=-0.291mV
  TPP=0.969s, TPV=0.175mV
  UPP=1.227s, UPV=0.085mV
  R-R Interval:0.625s [0.125s - 0.750s]
```

Fig. 9.13 Software simulation output

9.5.2   Data Display and Analysis

Thecoordinates of the ECG graph from each node are first transmitted through the
Internet and sent to the database. This process is done as long as there are new
incoming data. The continuity of the update process gives us a wide range of ECG
data classified by capture date. The real time charting library implemented in the web
visualization tool will manipulate this huge amount of data. So, the medical staff can
consult the latest incoming data at real time (a marker is implemented to show that
there are new incoming data) or it can re-consult previous data.

As it is shown in Fig. 9.16, the tool contains four main modules. The most important one is the ECG viewer itself with the capability to display the data from three leads
at the same time. The second module (window) displays the nodes which already
sent data to the database. The remaining utilities are dedicated for the information
related to the node (patient).

(a)
```
$.ajax({
            //Call the script to get new peak
            url: 'live-server-data.php',
            success: function(point) {
            var series = chart.series[0],

        // add the peak
            chart.series[0].addPoint(point, true, shift);

            // call it again after one second
              setTimeout(requestData, 1000);
                    },
                    cache: false
            });
```

(b)
```
    <?php include 'get_data.php'; ?>

        var dataFromPHP = [ <?php echo join($data, ',') ?> ] ;

    /* preprocess data */
    var newData = [];function pointExists(array, key) {
        for (var i = 0; i < array.length; i++) {
            if (array[i][0] == key) {
                return true
            };
        }
        return false;
    }


    /* 1) create only one point */
    for (var i = 0; i < dataFromPHP.length; i++) {
        if (!pointExists(newData, dataFromPHP[i][0])) {
            newData.push(dataFromPHP[i]);
        }
    }
```

Fig. 9.14 (a) Get live-data, (b) Get previous-data

Fig. 9.15 Multicore SoC system running snapshot

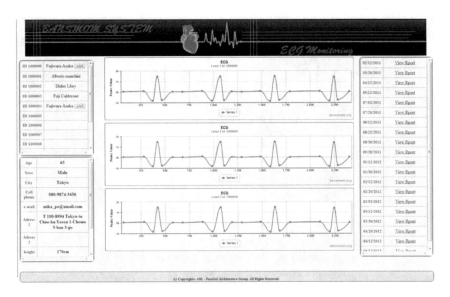

Fig. 9.16 Interactive RTI tool displaying ECG waves

Figure 9.14b illustrates the PHP script for consulting previous ECG data. In the case of visualizing live incoming data, an Ajax call is executed; this execution returns the last peaks that have been added to the database as shown in Fig. 9.14a.

9.6 System Hardware and Software Design Evaluation

9.6.1 Hardware Complexity

The system was designed in Verilog HDL. Figure 9.11 shows the block diagram of the prototyped system. The master module consists of the Altera Nios II core [14], four on-chip memories (used for raw ECG data storage, processor memory, shared memory and virtual external memory), an interrupt timer, a graphics LCD controller, a LED controller and a JTAG UART (used for connection with host PC). One PPD node (slave) consists of Nios II core, on-chip memory, and an interrupt timer.

The FIR filter module is generated by Altera MegaCore Function. The specification of this filter is as follows: filter step is 51, sample rate is 128, and cutoff frequency is from 5–15 Hz. Table 9.1 shows logic synthesis result. For example, one configured with 1-lead (only one PPD module), the logic utilization is about 14%, the total block memory bits is about 21%, and The total power dissipation is about 677 mW.

Table 9.1 Hardware complexity

System model	Logic utilization				MEM bits	Speed (MHz)	Power (mW)
	ALUTs	ALUTs	REG	Total (%)			
1-lead	9,769	16	11,669	14	1,207,312 (21%)	97.89	677.00
2-lead	17,169	32	21,297	26	1,810,384 (32%)	95.82	716.31
3-lead	24,592	48	30,947	38	2,413,840 (43%)	92.52	754.84
4-lead	32,047	64	40,566	50	3,016,976 (54%)	92.25	784.31

Table 9.2 Performance evaluation

Recode (No.)	Detected RR Interval (# of interval)	Failed Detection (# of interval)	Execution time (s)
16265	14	7 (50%)	6.787
16273	13	3 (23%)	6.959
16420	14	5 (36%)	6.791
16773	10	1 (10%)	6.511
16786	10	3 (30%)	6.524
17052	9	2 (22%)	6.182
18177	15	5 (33%)	8.316
18184	8	3 (38%)	4.860

9.6.2 Performance Evaluation

We used real sample data from PhysioBank data base [17] for testing the correctness and the accuracy of BANSMOM system. Table 9.2 shows the test results over various configurations. Figure 9.13 shows the simulation screen capture of the processing report. PPP, QPP, RPP, SPP, TPP and UPP mean P, Q, R, S, T and U Peaks. On average, the PPD algorithm achieves fair accuracy of about 69%.

9.7 Chapter Summary

Recent technological advances in wireless networking, microelectronics and the Internet allow computer and biomedical scientists to fundamentally modernize and change the way health care services are deployed.

Electrocardiography is a commonly used, non-invasive procedure for recording electrical changes in the heart. The record, which is called an electrocardiogram (ECG or EKG), shows the series of waves that relate to the electrical impulses which occur during each beat of the heart. The results are printed on paper or displayed on a monitor. The waves in a normal record are named P, Q, R, S, T, U and follow in alphabetical order.

MCSoCs are high performance devices that incorporate multiple building blocks from multiple sources. An embedded multicore SoC may contain general or special purpose fully programmed processors, co-processors, DSPs, dedicated hardware, memory blocks, etc. These MCSoC systems are becoming a common design alternative in portable devices because it is possible to manufacture a silicon chip including only necessary elements.

This chapter exploits the technology of parallel processing to process the electrocardiography computational kernels in parallel. The idea is to implement the traditional multi-lead bulky electrocardiogram on a programmable embedded multicore SoC which is small and more efficient. The presented solution paves the way for real-time processing diagnosis of heart-related diseases. Prototyping of Multicore SoC on FPGA involves building a functional system model that lets the designer evaluate various aspects of a design, and provide a realistic projection about the final product implementation. Having a prototype available provides more accurate power-performance evaluations. For example, prototypes make accurate area estimations feasible, as well as hardware complexity overhead and energy consumption measurements.

References

1. Y. Haga, A. Ben Abdallah, K. Kuroda, Embedded MCSoC architecture and period-peak detection (PPD) algorithm for ECG/EKG processing, in *The 19th Intelligent System Symposium (FAN 2009)*, (2009), pp. 298–303
2. A.D. Desai, T.S. Yaw, T. Yamazaki, A. Kaykha, S. Chun, V.F. Froelicher, Prognostic significance of quantitative qrs duration. Am. J. Med. **119**(7), 600–606 (2006)
3. G.M. Friesen, T.C. Jannett, M.A. Jadallah, S.L. Yates, S.R. Quintand, H.T. Nagle, A comparison of the noise sensitivity. IEEE Trans. Biomed. Eng. **37**(1), 85–89 (1990)
4. J. Pan, W.J. Tompkins, A real-time QRS detection algorithm. IEEE Trans. Biomed. Eng. **32**(3), 230–236 (1985)
5. M.G. Carey, Electrocardiographic predictors of sudden cardiac death. J. Cardiovasc. Nurs. **23**(2), 175–182 (2008)
6. Y. Jewajinda, P. Chongstitvatana, *Electrical Engineering/Electronics Computer Telecommunications and Information Technology (ECTI-CON) IEEE*, (2010), pp. 1050–1054
7. B.U. Kohler, C. Hennig, R. Orglmeister, The principles of software QRS detection. Eng. Med. Biol. Mag. IEEE **21**(1), 42–57 (2002)
8. F. Magrabi, H. Nigel, B.G. Celler, Web based longitudinal ECG monitoring, engineering in medicine and biology society, in *Proceedings of the 20th Annual International Conference of the IEEE*, (1998), pp. 1155–1158
9. J. Dong, J.-W. Zhang, H.-H. Zhu, L.-P. Wang, X. Liu, Z.-J. Li, Intelligent systems. IEEE **1**, 36–43 (2012)
10. A. Ben Abdallah, Y. Haga, K. Kuroda, An efficient algorithm and embedded multicore implementation for ECG analysis in multi-lead electrocardiogram records, in *IEEE Proceedings of the 39th International Conference on Parallel Processing Workshop*, (San Diego, 13–16 Sept 2010), pp. 99–103
11. A.B. Ahmed, Y. Kimezawa, A. Ben Abdallah, Towards smart health monitoring system for elderly people, in *IEEE Proceedings of the 4th International Conference on Awareness Science and Technology*, (2012), pp. 248–253
12. H. Yasuyoshi, A. Ben Abdallah, Architecture and design of application specific multicore SoC, graduation thesis, The Univ. of Aizu, Feb. 2010
13. Altera Design Software, http://www.altera.com/
14. Nios II Processor, http://www.altera.com/literature/lit-nio2.jsp
15. PHP Hypertext Preprocessor, http://www.php.net/
16. Mysql, http://www.mysql.com/
17. PhysioBank, http://www.physionet.org/physiobank/

Index

© Springer Nature Singapore Pte Ltd. 2017
A. Ben Abdallah, *Advanced Multicore Systems-On-Chip*,
DOI 10.1007/978-981-10-6092-2

Printed in the United States
By Bookmasters